Life Narratives, Creativity, and the Social in the Americas

American Frictions

Editors
Carsten Junker
Julia Roth
Darieck Scott

Volume 11

Life Narratives, Creativity, and the Social in the Americas

Edited by
Wilfried Raussert and Susana Rocha Teixeira

DE GRUYTER

ISBN 978-3-11-155022-0
e-ISBN (PDF) 978-3-11-155196-8
e-ISBN (EPUB) 978-3-11-155270-5
ISSN 2698-5349

Library of Congress Control Number: 2024943398

Bibliographic information published by the Deutsche Nationalbibliothek
The Deutsche Nationalbibliothek lists this publication in the Deutsche Nationalbibliografie; detailed bibliographic data are available on the Internet at http://dnb.dnb.de.

Typesetting: Integra Software Services Pvt. Ltd.
Printing and binding: CPI books GmbH, Leck

www.degruyter.com

Acknowledgments

For a book like this, acknowledgments must be paid to the contributors who have not only enriched us with their presence at our two-day intensive workshop, "Life Writing, Creativity, and the Social in the Americas" but also prompted us to put together this book due to the enriching discussions about life writing, Hemispheric American Studies, the connections between creativity and the social world and between self and collective narratives at the Center for InterAmerican Studies at Bielefeld University in June 2022.

We are deeply grateful and learned much from Mita Banerjee, Sofie Behluli, Anna Ferrari, Atalie Gerhard, Astrid Haas, Alfred Hornung, Frank Mehring, Ines E. Veauthier, and Sabrina Völz during our presentations, discussions, and revision processes.

The impetus for writing this book was also possible due to the funding of our research and workshop by American Studies and CIAS, Bielefeld University.

We received very careful and eagle eye observational help from Alexa Kenter during all stages of preparing and editing this publication – we owe you many thanks, Alexa. Lastly, we would like to thank Sarah-Lena Essifi and Dafni Tzouna for the assistance in the organization of the workshop. And we would like to thank the De Gruyter team Julie Miess and Anne Stroka for a very smooth publication process.

Wilfried Raussert & Susana Rocha Teixeira

https://doi.org/10.1515/9783111551968-202

Contents

Wilfried Raussert and Susana Rocha Teixeira

Introduction: Life Narratives, Creativity, and the Social in the Americas

It appears that life narratives have been *en vogue* for quite some time now. While scholars such as Julie Rak or Leigh Gilmore diagnosed a "memoir boom" in recent decades (Rak 2013; Gilmore 2001; Douglas and Poletti 2016, 44; Novak 2017, 1), for Susanna Egan and Gabriele Helms, "[t]he last quarter of the twentieth century has been a profoundly autobiographical age with personal narrative providing an important lens both on to history and on to the contemporary world" (Egan and Helms 2004, 216). At least in the last three decades, the number of (auto-)biographies, memoirs, and other life writing subgenres published and on bestseller lists seem to point to what appears to be an insatiable hunger for (the) lives (of others) among readers and within the publishing and print industry (Rak 2013). Similar holds true for other forms of life narratives. The commercial success of the Netflix documentary series *Meghan and Harry* (2022) that preceded the Duke of Sussexes memoir *Spare* (2023), or even the success of the *Tiger King* Netflix documentary (2020) that accompanied the beginning of the COVID-19-lockdown era appear to be cases in point. In December 2022, Sarah Shaffi in *The Guardian* may have been wondering whether readers and producers of memoirs were losing interest in celebrity memoirs, however the previous examples appear to point to an ongoing interest in life narratives of the rich, the famous, and the extraordinary.

The interest in life narratives is also discernible in academia and the academic community as a growing number of journals, groups, and book series on the subject attest.[1] While for a long stretch of time the classical forms of (auto-)biography as accounts of human lives were considered the only object of study (Hornung 2013b, x), albeit not because of their literariness or creativity, which is an aspect that was neglected for a long time (cf. Cardell 2014, 14; Olney 1980, 4), the latter half of the twentieth century showed several new developments in the study of autobi-

1 This includes, for example, Wilfrid Laurier University Press's Life Writing series, edited by Marlene Kadar and Sonja Boon, The Palgrave Studies in Life Writing series, edited by Clare Brant and Max Saunders, The Wisconsin Studies in Autobiography series, edited by William L. Andrews, the Routledge Auto/Biography Studies series, edited by Ricia A. Chansky, Bloomsbury's New Directions in Life Narrative series, edited by Kate Douglas, John David Zuern, and Anna Poletti, Brill's Egodocuments and History Series, edited by Arianne Baggerman, Rudolf M. Dekker, and Michael Masuch, and Brill's Biography Studies series, edited by Hans Renders, as well as third-party funded projects such as the Freie Universität Berlin's DFG Research Group 530 Self-Narratives in Transcultural Perspective (2004–2012).

https://doi.org/10.1515/9783111551968-001

ography and biography. In the last decades, research has continuously added other forms of written expressions of one's own life and the lives of others such as internet blogs or the diary (cf., e.g., Smith and Watson 2010) which had been neglected before due to being presumably private texts or their lack of literariness or "artistic intent" (Cardell 2014, 13). This increased interest resulted in new perspectives on this traditional genre and led to the usage of the more inclusive term life writing to describe heterogeneous forms of written expressions of one's own life and the lives of others (cf. Smith and Watson 2010). Life writing not only takes many different forms but has also been increasingly studied from a variety of perspectives, ranging from, e.g., literary Studies to history, philosophy, and psychology (Farr and Ruggiero 2022, 2).

Probably since the beginning of human life, new forms of media and their possibilities for human self-expression contributed to novel forms of representing and narrating oneself and others. We find the impulse of life writing in various textual genres such as autobiography, slave narrative, the novel, or the autobiographical essay to name but a few. Furthermore, internet blogs, social media, portraiture, (documentary) films, dance, and so on not only challenge and expand what life writing means but also point to the limitations of the focus on writing (Smith and Watson 2010; Hornung 2013b, x; Maguire 2018, 9; Poletti 2020). In response, some scholars suggested concepts such as automediality instead of autobiography since the former takes different media and forms of (self-) representation into consideration as well as the interplay between them (Dünne and Moser 2008; Renov 2004).

Automediality yet excludes lives that are narrated or expressed by others, which is why we as editors prefer the term life narrative; the latter allows us to highlight the rich spectrum of numerous forms of narrating oneself and others (Douglas and Poletti 2016, 11).[2] This term also makes it possible to take the increasing presence of intermedial expressions as part of life narratives into consideration, to give voice to a growing concern with aesthetics in and of life narratives, and to illustrate boundary-crossing processes at work. "Fine-tuned usage of intermedial intersections of meaning-making possibilities," Nassim Winnie Balestrini and Ina Bergmann point out, "also facilitates performative life writing in the sense that the autobiographical trajectory is directed towards the artistry of the depiction as much as or possibly more than the ostensible completeness of narrating a life" (2018, 2). Building upon these discussions, our edited volume explores the intersections of life narratives, the social, creativity, and different forms of media to narrate and (re)present the self to see in which way these expressions

2 Critics such as Emma Maguire pointed to the issues associated with the term life narrative (2018, 20–21).

offer (new) means of (self-) representation (cf. Poletti and Rak 2014). The contributors of our edited volume resort to a variety of concepts or terms, which itself attests to the variety of different perspectives in academia and elsewhere on the subject of narrating oneself and others.[3]

Fact versus fiction

Resorting to life narrative as a comprehensive umbrella term makes it possible to challenge distinctions between fact and fiction and related claims to veracity, authority, and authenticity that Lejeune's autobiographical pact and scholars such as Couser associated with non-fiction (sub-)genres such as biography, autobiography, or diary (cf. Couser 2021, 7, 13, 68; Novak 2017, 13; Rak 2009; McLennan 2013, 10–11). According to this view, the latter is heavily based upon lived experience. This experience might be shaped by context, it might be mediated, reflected, or imaginatively transformed like other forms of cultural or artistic production. Despite the presence of imaginative storytelling and fictionalizing elements, according to this 'school' of thought, the presumably non-fiction (sub-)genres contain an air that seems always a step closer to real lives and their baggage of memory, trauma, reality, hope, and vision. Different from fiction writing, these forms of life narratives turn lived experience into narrative(s) that wish to keep the thread between lived life and storytelling intact, most if not all the time. It is important to note, though, that while there is a strong mimetic impulse guiding conventional, nonfictional forms of life narratives, all forms of cultural production are creative acts. Creativity in the context of life narratives nourishes the act of narrating and propels among others the desire to link individual life stories with larger stories of social embeddedness, conditioning, and transformation thus pushing new forms of historiography and other forms of nonfictional writing. Accordingly, the creative impulse fuses individual and collective experience with a larger understanding of the social including the latter's local and global embeddedness.

Increasingly, research has complicated the seemingly clear-cut distinction between the presumably factual, fictional, and literary or aesthetic value. Some critics, for example, pointed to the power of auto-biographical and other forms of fiction to reveal particular truths that might remain hidden in nonfictional gen-

---

3 For equally convincing reasons, it appears that, e.g., Maguire has a diametrically opposed view (2018, 21). Due to the many forms of narrating oneself and others that are possible, Couser concludes: "What we call *life writing* we should really call *life representation*" (2021, 13). For him, however, "[t]he key criterion is that the subjects are not invented" (13).

res, for example regarding the subject's real sentiments (cf. Novak 2017, 7; Novak and Dhúill 2022, 22; Dix 2018, 6). This echoes a view voiced by Mark Twain that the "real life" of a person is hard to capture in biographies, which are "but the clothes and buttons of the man" while "the biography of the man himself cannot be written" (Twain 2010, 220–21; Couser 1989, 106). Nonfictional genres, too, have their limitations and difficulties regarding capturing a person's 'real' life. The fact that Mark Twain's autobiography was published posthumously and that he didn't use his real name, Samuel Clemens, further challenge claims to veracity allegedly offered by this nonfictional genre (cf. Couser 1989, 11). At the same time, Twain's autobiography appears to confirm Mita Banerjee's claim that it "can be seen as a paradigm case of the writing of autobiography as such . . . encapsulat[ing, eds.] virtually all of the issues which have been related to writing an autobiography" (2019, 1659).

Lies, fabrications, and untruths further complicate the distinction between presumably factual and fictional genres (cf. Vice 2014). This holds true for literary hoaxes, where authors, for example, assume a different racial or sexual identity, often for their own benefit (Couser 2021, 68). Infamous confidence men and women such as Russian born Anna Sorokin who managed to enter New York high society by pretending to be a rich heiress named Anna Delvy, American Congressman George Santos, and Rachel Dolezal, who (in)famously pretended to be Black, also blur the lines between fact and fiction in their invented life stories. Similar holds even true for individual life events such as Ernest Hemingway's or William Faulkner's claims regarding their military service during World War I, which were exaggerated or fabrications, respectively (cf. Coté 2002; Davis 2020, 144). However, although Novak and Ní Dhúill pointed out that "biographical fiction and factual biography can['t, eds.] simply be conflated – they clearly operate under different truth contracts" (2022, 8), they also claimed that "life narratives are ideological constructs, more or less carefully crafted for consumption by specific audiences for specific purposes and with reference to existing conventions of genre" (2).

Some scholars pointed to the blurred lines between the novel and other forms of fiction and presumably non-fictional forms of writing such as the memoir. One example is the "recent explosion of experimentation in life-writing" (Novak 2017, 2) that includes forms of narrating lives such as "'autofiction,' 'biofiction,' 'auto/biografiction,' or 'autobiographical non-fiction novel'" that challenge genre conventions (4) and distinctions between fact and fiction. However, this fuzziness regarding fact and fiction in the context of life narratives is not new. Scholars have highlighted for some time the joint development auto-biography and the (realist) novel in the nineteenth century, their shared interest in character development, and their similar narrative strategies (Novak 2017, 5; Couser

2021, 13), which some critics attributed to the residual importance of the realist novel (cf. Novak 2017, 5–6). Consequently, it doesn't appear surprising that non-fictional forms of writing such as autobiography or biography and even 'raw' and 'unmediated' written expressions such as the diary are like their fictional counterparts products of creativity that like other forms of creative writing tend to follow their expected genre conventions and to resort to narrative strategies, literary devices, and methods to produce particular effects such as a reality or authenticity effect that creates the notion among readers of being given a particular unmediated insight into reality or (past) truths or facts (Cardell 2014, 5; Nünning 2022, 29; Novak 2017, 6–9; Gudmundsdottir 2017, 54, 164).[4] Works such as James Weldon Johnson's novel *The Autobiography of an Ex-Colored Man* (1912/1927) play with this messiness between fact and fiction by blurring the lines between the author and the main character, Bildungsroman, and autobiography or the distinction between the factual and fictional altogether, which allows to expose other alleged objective truths as fictions, for example regarding differences between the races, views that were still prevalent in the United States of America when Johnson's novel was published. Furthermore, by pointing to established genre conventions and the concomitant possibilities for character development associated with both Bildungsroman and autobiography, Johnson's *Autobiography* highlights the limitations Black Americans faced in American society and regarding different forms of cultural or artistic production (Hinrichsen 2015, 177; Morrissette 2017, 11).

Re-producing authentic lives

In roughly the last two decades, the internet, and especially websites, blogs, or social media platforms such as *Facebook*, founded 2004, *Twitter* (now: *X*), founded 2006, or *Instagram*, released in 2010, offered new opportunities for creating a reality or authenticity effect and narrating or presenting authentic selves not only for the rich and the famous, but also for the average Joe or Jane. This appears to make this form of narrating or presenting the self and others even more democratic than autobiography or memoir, genres which were already considered to be of democratic essence by critics such as W.D. Howells, and thus according to some, a particular American genre (cf. Howells 1909; Couser 2021, 20, 32, 35; Danielewicz 2018, 3; McLennan 2013, 8, 30, 32; Poletti 2020, 176; Poletti 2014). Similar

4 Cf. Couser (2021, 17) on creativity in Thoreau. For a more detailed discussion of autofiction/ biofiction, cf. Dix (2018); Effe and Lawlor (2022); Novak and Ní Dhúill (2022).

holds true for reality TV shows, (Maguire 2018, 11–12; Danielewicz 2018, 10; Helms 2005, 43–63; Thiemann 2019, 796) which, for example, Helms described as "auto/biographical practices and . . . part of a long tradition of such practices in a wide range of media" (Helms 2005, 46). Despite appearances, reality TV shows as well as other formats that seem to offer nonfictional life narratives do not provide an unmediated access to a particular real or authentic self. Instead, they are cultural products, artistic, and cultural expressions and the result of creativity, hard work, genre conventions, and the market (Helms 2005, 45) that appears to have demanded and offered the confessional and realist mode in the last decades (Cardell 2014, 17).

As David Foster Wallace shows in "How Tracy Austin Broke My Heart," which can be seen as an exemplary review of a "mass-market book – the sports-star-'with' – somebody autobiography" (2006, 141), nonfictional life narratives do neither necessarily convey particular truths nor record exemplary and unique lives. Instead, as this example shows, nonfictional life narratives can appear generic, primarily geared toward the market and profit and devoid of real content or meaning. The actual tragic life of Tracy Austin seems to be a particular suited example because her autobiography doesn't appear to attempt to convey the real suffering she must have gone through due to the various tragic and dramatic life events that turned her from a tennis prodigy to an injured ex-tennis player. Instead, according to David Foster Wallace, her autobiography *Beyond Center Court: My Story* appears to be a badly (ghost-) written collection of generic conventions, truisms, and platitudes. Although "we want to know how it *feels*" (2006, 143),

> [i]t's almost as if Tracy Austin has structured her own sense of her life and career to accord with the formulas of the generic sports bio. We've got the sensitive and doting mother, the kindly dad, the mischievous siblings who treat famous Tracy like just another kid. We've got the ingenue heroine whose innocence is eroded by experience and transcended through sheer grit; we've got the gruff but tender-hearted coach and the coolly skeptical veterans who finally accept the heroine. We've got the wicked, backstabbing rival (in Pam Shriver, who receives the book's only unfulsome mention). We even get the myth-requisite humble roots. (145)

In the last decades, critics have argued that the notion of authenticity is problematic itself, because "the self [itself, eds.] is a work of art and a product of invention" (Cardell 2014, 29). As Eakin puts it: "our sense of continuous identity is a fiction, the primary fiction of all self-narration" (Eakin 1999, 93; Farr and Ruggiero 2022, 4–5; Greenblatt 1980, 256–57). Following this view, life writing cannot (accurately) record or present someone's life or being. Instead, and reminiscent of Ricoeur's notion of narrative identity, the narrative (creatively) produces that person's identity and life, and thus ultimately that person or individual (Maguire 2018, 12–13). Furthermore, notions of a stable, whole, authentic self or identity have been challenged for quite some time, e.g., by psychoanalysis à la Sigmund Freud and in poststructuralist thought, which decentered and fragmented the subject. Deconstructivist thought as advocated, for example, by Roland Barthes, famously challenged related notions about the author, who then no longer necessarily preceded and produced the text, but rather became part of it. To make matters worse, current research on bacteria and the human brain appears to suggest that bacteria affect the way humans think and behave (Sender, Fuchs, and Milo 2016; Dinan et al. 2015; Skonieczna-Żydecka et al. 2018; Gao et al. 2019). This has raised serious ontological questions regarding human nature, agency, and consequently life writing. If bacteria shape our thoughts and feelings, then who is actually in control of our brains, ourselves, and (creatively) producing life narratives, and how do these and other non-human life narratives look like?[5]

Life narratives and the social

Life narratives not only represent or produce individual selves, but they are also entangled with the social. Kyle Cardell claims that "contemporary diaries make visible the intimate and the personal," but that they also "blur and destabilize conventional boundaries between public and private" (2014, 3).[6] For Cardell, diaries are "a popular form . . . through which key social discourses are played out and made visible" (2014, 4). We believe that this also applies to other forms of narrating oneself and others. Life narratives, in our view, reflect, critique, affirm, and negotiate social practices and change, and in doing so not only reflect, but also participate in national and transnational conversations and contribute to social developments (cf. Kleinreesink 2017, 14, 125; Danielewicz 2018, 1, 3, 8). Life writing scholarship explores these and related issues as the interest in the social increasingly grows (Douglas and Poletti 2016, 11, 47).

Echoing developments in other areas of academia and elsewhere in the second half of the twentieth century, following, for example poststructuralist and feminist thought, new fields such as Postmodern and Postcolonial Literary and Cultural Studies contributed to new perspectives and objects of study regarding life narratives. James Olney's publication *Autobiography: Essays Theoretical and Critical* in

5 For further discussions on non-humans or posthumanism, cf. Batzke, Garrido, and Hess (2021); Novak (2017, 14); Hornung (2013b, x); Lanzendörfer (2013); Couser (2021, 13).

6 Douglas and Poletti show by resorting to current life writing scholarship, that, e.g., sharing of intimate details is not limited to social media. For example, diaries or letters were once "read aloud to family members, shared and reported on" (Douglas and Poletti 2016, 19, 21–22).

1980 is an example for how the autobiographical canon (traditionally considered as an exclusive product of Western male culture) was expanded to investigate and therefore include previously neglected forms of self-representation and narration (for example by African American writers) during that period, and how the central position of the white, heterosexual, healthy, Western male subject increasingly got challenged and decentered in the area of life narratives (Jones 2007; Gilmore 2001; Rak 2013; Smith and Watson 2001, 2006, 2010; Douglas and Poletti 2016, 8).

This change of perspective and expansion of the life narratives canon continued in the late 1980s and beginning of the 1990s with works such as Paul John Eakin's *Fictions of Autobiography: Studies in the Art of Self-Invention* (1985) and Sidonie Smith and Julia Watson's influential *De/Colonizing the Subject: The Politics of Gender in Women's Autobiography* (1992). Since then, the interdisciplinary study of life writing has seen a shift that has increasingly opened the floor to the expressions of women, who added female perspectives, personalities, notions of the self, as well as (sub-)genres (such as the diary) that had been neglected before (cf. D'Amore 2012, 2; Kadar 2005, 2; Hart 2004; Novak and Ní Dhúill 2022, 8; Olney 1980). Similar holds true for LGTBIQ+ persons (cf. Novak and Ní Dhúill 2022, 16), (postcolonial) migrants, and ethnic minorities; "personal narratives no longer depend [exclusively] on speakers belonging to dominant social groups but emerge . . . from minority positions, cultivate the value of undervalued experiences, and risk distinctly intimate subject matter" (Egan and Helms 2004, 216; Douglas and Poletti 2016, 11). As Douglas and Poletti put it: "life narrative has moved from being a cultural form associated with the lives of 'great men' to being a dynamic and influential means for people and communities to write themselves into culture and history" (2016, 8).

Following the somatic turn in academia in the last decades, scholarly interest in the body grew in the field of narrating lives. This includes, e.g., the landmark study by John Paul Eakin, *How Our Lives Become Stories. Making Selves* (1999) as well as numerous works by Thomas Couser such as his *Recovering Bodies: Illness, Disability, and Life Writing* (1997), or *Signifying Bodies: Disability in Contemporary Life Writing* (2009). The study of bodies in the field of life narratives does not only revolve around healthy male and female bodies, which conform to hegemonic and heteronormative standards. On the contrary, as accounts during the AIDS pandemic or of cancerous bodies show, it also very much includes narratives about disabled, ill, disfigured, or deceased bodies, which played a significant role in the fight for 'minority' rights (Couser 2021, 5ff) and contributed to transformations of the social. As it has become clear, life narratives are not so much concerned with mimetic representations of societal issues (Kleinreesink 2017, 14). For example, the prevalence of certain disability or illness narratives in the literary market or within cultural products does not necessarily have something to do with the prevalence of a particular illness within society. If this were the case, more life narratives should be written about the female lung cancer than about breast cancer, as Couser pointed out (2021, 28). Similar holds true for illnesses such as heart disease or diabetes which are much more common than, for example, HIV, but receive much less attention in the field of life narratives (2021, 28). These life narratives are often what Couser considers "*some body* memoirs," this means memoirs that 'nobodies' (17–19), aka, average people, produce about "a *particular* body" (18). These forms of – sometimes self-published – life narratives that tend to deal with childhood, physical or mental impairments, further exemplify the basically democratic nature of the genre and are manifestations of its continued democratization (18, 32). They also show that the creative impulse is often stronger than mimetic desire behind narratives.

Although these narratives might give voice to previously neglected groups, many of these life narratives can be categorized as "narratives of overcoming" (29) or "narrative[s] of emancipation," which is problematic for a number of reasons, one being that some activists criticized this as "inspiration porn," because it is very rare in real life to overcome disability (29–30). At the same time, the increased existence and visibility of alternative voices, was not only made possible by social change, but also contributed to it (19).

Rethinking the social, we as editors maintain, is an important aspect that should be emphasized more in studies of life narratives. As Thomas Couser puts it, life writing "acts in and on history in significant ways" (2021, Foreword n. pag.). A hemispherical approach to the social in the Americas serves as an ideal starting point to challenge the concept in new ways questioning the postmodern assumption that the social is dead (Kaltmeier and Raussert 2019, 3). In the twenty-first century, Bruno Latour has been among the major voices articulating the crisis of the social. Stressing the "many . . . contradictory cartographies of the social" (2005, 34), he emphasizes the interaction between human and nonhuman actors and raises new questions about interconnections and interrelationships within and between plural societies. Ethnic, environmental, religious, and women movements have been a major factor in breaking with traditional concepts of the social and inducing change as well provoking right-wing backlash. Due to migration and displacement, post-territorial narrations of the social have emerged that go far beyond nation-state projects. Concepts such as family, community, society, state, and nation are too narrow in range and specified to do justice to the "fluidity of the social" (Bauman 2000, 25).

Following scholars that have highlighted the complex interrelations between life narratives and the social, we aim at exploring how social context, developments, and transformations, which include crises such as illness, identity crisis, racism, and sexism challenge and redefine how we (re-)present ourselves. We explore, for example, how life stories of or about people who have historically been forced into subordinate positions, exploited, and oppressed contributed to new social developments or how social developments contributed to new forms of (re-)presenting the self and others and other related issues[7] as it can be seen in social movements such as *Black Lives Matter* or *MeToo*. Recently, Disability Studies and Life Narrative Studies have joined hands in pointing out that certain forms of life narratives "enact[ ] and sustain[ ] human relationships" (Couser 2021, Foreword n. pag.). Disability memoirs have emerged with the force of the disability rights movement from the 1980s onward with its landmark achievement in the United States, in 1990, of the Americans with Disabilities Act (ADA), a civil rights law that banned discrimination against people with disabilities (20). These narratives have underscored life writing's tremendous capacity to make the unheard heard and the invisible seen.

Life narratives in the Americas

Building upon American Studies, Transnational and Hemispheric American Studies, and Inter-American Studies (Fisher Fishkin 2005; Hornung 2007; Thies and Raab 2009; Fluck, Pease, and Rowe 2011; Lenz 2011a, 2011b; Hebel 2012; Davis 2013; Raussert 2014) this book challenges the often limited, limiting, exclusive, and excluding focus on the United States that is discernible in the field of American Studies (cf. Couser 1989; Eakin 1991; Sayre 1978) by exploring life narratives in the Americas. In doing so, our book takes previous discussions about life narratives in the Americas (Chansky 2017; Graham and Lucero 2020; Ortiz-Vilarelle 2021) as starting point to systematically explore the myriad connections between life narratives, creativity, and the social in the Americas, and the role of life narrative genres as expanding and changing for reflections on, and inventions of the social from a variety of perspectives. Looking at life narratives from the perspectives of Hemispheric American Studies it is important to include discourses beyond hegemonic U.S. paradigms which address various contact zones and forms of (im)mobility in the Americas. For example, Smith and Watson argue that in life-writing discourses, "[i]mmigrant narratives . . . have long been sites through which formerly marginal and displaced ethnic and racialized subjects explore terms of their cultural identities and their diasporic and transnational allegiances" (Smith and Watson 2010, 156). Moreover, Rosemary Marangoly George argues for a distinct 'Immigrant Genre' – defined according to "political and ideological contents rather than by formal attributes" – "which is marked by a disregard for national

7 Cf. Jensen and Jolly (2014) for life narratives and human rights. Cf. Schaffer and Smith (2004).

schemes, the use of multigenerational cast of characters . . . [and] a curiously detached reading of the experience of 'homelessness' . . ." (George 1999, 171).

Through the lens of life writing within a Caribbean diaspora context, Lisa R. Brown argues, that modes of life writing have long served as an important function in Caribbean literature: "Prose narratives, travel narratives, fictional autobiographies, . . . slave narratives, community histories, biographies, autobiographical poems, [and] memoirs . . . all represent points on the vast continuum of use through which Caribbean creative writers and 'real people' alike have made self-representation" (Brown 2011, 276). However,

> [t]he use and function of each mode has varied over time from the earliest writings by colonial subjects transplanted to the Caribbean either as slaves, colonial masters, or indentured labourers, to the writings of the first wave of nationalist writers who migrated to England, to the most recent writers residing in the region or in the diaspora. (276)

Each of these modes may be viewed as having a particular "performative function" (276) which, pertaining to the most recent writers, seems to have significantly shifted "from a very political nation-building enterprise to a more personal life-saving project" (277). Particularly the concepts of "performativity, positionality, and relationality" (Smith and Watson 2010, 214) have contributed to theorizing and contextualizing this immense variety of (auto-)biographical forms of self-expression.

Among the most prominent topics in Hemispheric American Studies, Transnational American Studies (Davis 2013; Fisher Fishkin 2005; Fluck, Pease, and Rowe 2011; Hebel 2012), and Transcultural American Studies (Hornung 2007; Lenz 2011b; Raussert and Isensee 2008) have been questions of identity, belonging, and affiliations particularly regarding processes of globalization. This trend is also recognizable in Canadian and in Postcolonial Caribbean Cultural and Literary Studies (Brydon and Dvorák 2012; Dobson 2009; Kamboureli and Roy 2007). Citizenship "has become deeply self-different and has lost its traditional all-comprising claims of a public identity . . . and its basis in a closed concept of national territory and unified national culture that was seen as providing an abiding sense of belonging" (Lenz 2011a, 5). New transnational and transcultural perspectives are thus not only needed to explain (post)modern phenomena but, in fact, may open up innovative and intriguing new ways of thinking about human ways of narrating issues of identity, belonging, and affiliations. As Günter Lenz states:

> The decentering and deconstructing forces of the transnational conceptions and practices of citizenship threaten the traditional sense of home, of belonging, of community, but they also open up new options for forging more complex, multifaceted, and more experimental and localized visions of communal political and cultural practices in the globalizing world. (7)

Life writing plays a key role in this endeavor.

Life narratives and creativity

Although they are still often equated with their representational function, nonfictional life narratives have a lot of creative potential. This includes, for example, the creative interplay between fact and fiction in life narratives, creativity that emerges from the interplay of different forms of media, new connections between self and society, self and environment, and regarding alternatives to representational reflections and constructions of the social. For example, for Jocelyn Stitt, who is referring here to postcolonial writers "[t]he capaciousness of life-writing allows . . . authors to refashion the genre to create hybrid or creolized forms that can be understood as archives of experience and memory" (2017, 180). She further elaborates that "genre and form are reworked through the use of fragmented stories, combining multiple genres in one text, such as letters, legal documents, maps, photographs, drawings and personal narrative" (180). It appears, for example, that one of the reasons research in the field is so dynamic stems from the fact that it understands life, art, and criticism as interconnected rather than separate (Banerjee et al. 2011; Hornung 2009a, 2009b, 2013). Life writing has "developed into a privileged form of self-expression and into a major field of scholarship" (Hornung 2009a, 536). As such, it is a field that continues to visibly expand and include previously neglected texts and other audio-visual narrative genres. Consequently, this field is in persistent need of revaluation in terms of definition and methodology as concepts of self and narrative are never fixed. Our book explores the variety of forms of creativity that encompass but also go beyond the conventional notions associated with cultural and artistic production in the context of life narratives. At the same time, it is important to keep in mind that the creative acts of and within life narratives are strongly embedded in individual as well as collective experiences which go beyond representational and constructivist discourses.

A close look at the intersections of experience, creativity, and the social within life writing characterizes our approach to one of the most prominent cultural expressions blossoming in our times. Especially in light of colonial origins and Puritan notions of a new millennium, even the writing of personal diaries had a larger social claim as part of a new historiography. Early colonial texts fulfilled the function of a documentation of the new world, at the same time a reflection on the encounter with this unknown world. From the beginning, social as well as individual self-understanding were closely intertwined in colonial text production. This may explain why there is a very pronounced autobiographical impulse in the literatures of the Americas, but especially in the literatures of the United States to this day (cf. Couser 2012; Eakin 1991). Certainly, there are periods in which the autobiographical impulse is more prominent; nevertheless, it is evident that the idea that individual writing is always political and carries social implications, if not explicit messages, has been at the core of American writing.

The creative impulse in life narratives can be seen, on the one hand, in the creation of new social models of charity and coexistence, but also repeatedly in the texts pushing various political, economic, social, and cultural independence movements that have characterized the history of the Americas and drawn on ideas of liberation, autonomy, and self-creation. Separation from the old colonizing world demanded concepts and ideas that could promise and express difference. Ultimately, autobiographical narratives, as part of producing a new history, have always been a balancing act between documentation, reflection, and creation. For these reasons, this book intends to see life narratives from the Americas in relation to the social and the creative to explore the important bridges between autobiographical impulse and the larger world.

Creative intervention, translation, and interpretation of lived lives into narrative shape synthetic and conflictive relationships between experience and imagination in the act of narrating lives. The power of life narratives lies precisely in their capacity to explore these dynamic relationships analytically and creatively and add alternative stories to local, national, and global historiographies. Perhaps more powerful than other forms of cultural or artistic expression, life narratives pursue and push an intense dialogue from the production to the reception of narrative(s) between self and other. Life narratives do not only attempt to make sense of experience, life, and larger history but are efforts to connect self and other simply by inviting readers/viewers/listeners to enter life stories intimately based on personal experience and memory. As Martina Wagner-Egelhaaf highlights, "[t]he new awareness of the 'real' and of people's 'real-life experience' . . . is not only due to the rapid developments in media and especially digital technology but also to the inexorable process of globalization which threatens to level out diversity and heterogeneity" (2019, 4). We need to take a close look at the "specific experientiality" of the selves of life narratives to keep track of lived difference, heterogeneity, and dissidence in times of homogenizing global discourses (Fludernik 1996, 6).

With a nod to the French philosopher Francois Lyotard, we maintain that life narratives bear the potential to create the small stories that should correct, relativize if not replace the grander narratives of history, religion, and philosophy in postmodern and contemporary times. In our intention to explore the congruences, tensions, and exchanges between experience, creativity, and the social, we look at the role that creativity plays on a textual and visual level but also how creativity is employed by the protagonists in life narratives to manage oppression, crisis, displacement, alienation, disease; we investigate how creativity is used by the protagonists to shape their environment, reflect the social, accept, oppose, and change life circumstances.

We maintain that the current boom of life narratives shows a heightened awareness of the social significance of acting creatively. Calling for new directions in studying cultural creativity as a socio-cultural phenomenon, leading scholars in the field of Creativity Studies point out that "creativity will become a necessity for the dignity and survival of the human species." This assumption rings true for communities in the past as well as the present. As the manifesto highlights, "creativity takes the form of action or activity" and occurs "in a given symbolic, social-institutional, and material context" (Glaveanu et al. 2019, 3). With a nod to these recent Creativity Studies developments, creativity is best defined as "a form of action in and on the world, performed in relation to others, and leading to continuous renewal of culture" (Glaveanu 2018, 84). This implies meaningful novelty in thought and action. Life narratives consciously and intentionally embrace and advocate both. Thus, life narratives become expressions of an instrumental function of art, expressed in the creation of new experience. "The first great consideration," says the American pragmatist philosopher John Dewey, "is that life goes on in an environment, not merely in it, but because of it, through interaction with it" (1981, 535). Art and culture do not arise in an autonomous space, but as a result of the interaction of the individual with their environment and their social circle.

Likewise, Dewey sees experience as always involving a larger social context, which he expresses in the following: "Experience in the degree in which it is experience is heightened vitality . . .; at its height it signifies complete interpenetration of self and the world of objects and events" (540). As he concludes: "[e]xperience is the result . . . of that interaction of organism and environment which, when it is carried to the full is a transformation of interaction into participation and communication" (543). Life narratives at their best are precisely such acts of transformation inviting further social reflection and change. These acts of transformation may include technological inventions, artistic creations, the creation of ideas for social enhancement, concepts for communal living, creation in everyday life culture, new directions in education, and environmental interaction in response to the respective condition of the social.

Methodologically, our book progresses in the three sections. While all contributions address the key concerns life narrative, creativity, and the social world as well as their entanglements, each section pursues specific priorities. The contributions by Alfred Hornung, Frank Mehring, and Susana Rocha Teixeira in section I Life Narratives play close attention to genre conventions and genre innovations and thus explore life narratives as a constantly developing and changing genre in transnational American contexts. Section II Creativity explores the role that creativity plays for and within the production of life stories as boundary-crossing impulse in aesthetic, narrative, and social terms. The contributions by Wilfried

Raussert, Astrid Haas, Ines Veauthier, and Sofie Behluli also highlight the creative uses of life narratives in social and historical contexts in the Americas. The book concludes with section III The Social. The contributions by Atalie Gerhard, Sabrina Völz, Mita Banerjee, and Anna Ferrari reflect upon ways in which life narratives explore, reflect, and create the social world and the ways in which changing personal, social, and historical crises shape the thematic concerns within life narratives and their forms of representation. By resorting to this triad structure we strongly believe to have created analytic clusters which allow us to see the rich spectrum and the aesthetic and social complexities of life narratives in and from the Americas.

Works cited

Balestrini, Nassim Winnie, and Ina Bergmann, eds. 2018. "Intermediality, Life Writing, and American Studies: A Brief Introduction." In *Intermediality, Life Writing, and American Studies*, 1–8. Berlin: de Gruyter.

Banerjee, Mita, Carmen Birkle, Bärbel Höttges, Manfred Siebald, and Nicole Waller. 2011. *Living American Studies*. Heidelberg: Winter.

Banerjee, Mita. 2019. "Mark Twain: Autobiography of Mark Twain (1870–1910)." In *Handbook of Autobiography/Autofiction*, ed. Martina Wagner-Egelhaff, 1659–1671. Berlin and Boston: De Gruyter.

Batzke Ina, Lea Espinoza Garrido, and Linda M. Hess. 2021. *Life Writing in the Posthuman Anthropocene*. Cham: Palgrave Macmillan.

Bauman, Zygmunt. 2000. *Liquid Modernity*. Cambridge: Polite Press.

Boldrini, Lucia, and Julia Novak. 2017. *Experiments in Life-Writing. Intersections of Auto/Biography and Fiction*. Cham: Palgrave Macmillan.

Brown, Lisa R. 2011. "Caribbean Life-Writing and Performative Liberation: Beyond the Boundaries." In *The Routledge Companion to Anglophone Caribbean Literature*, ed. Michael A. Bucknor and Alison Donnell, 276–84. London: Routledge.

Brydon, Diana, and Marta Dvorák, eds. 2012. *Crosstalk. Canadian and Global Imaginaries in Dialogue*. Waterloo: Wilfrid Laurier University Press.

Cardell, Kylie. 2014. *Dear World: Contemporary Uses of the Diary*. Wisconsin: University of Wisconsin Press.

Chansky, Ricia Anne, ed. 2017. *Auto/Biography across the Americas. Transnational Themes in Life Writing*. New York: Routledge.

Coté, William E. 2002. "Correspondent or Warrior? Hemingway's Murky World War II 'Combat' Experience." *Hemingway Review* 22, no. 1 (Fall 2002): 88–104.

Couser, G. Thomas. 1989. *Altered Egos. Authority in American Autobiography*. Oxford: Oxford University Press.

Couser, G. Thomas. 2012. *Memoir. An Introduction*. Oxford: Oxford University Press.

Couser, G. Thomas. 2021. *The Work of Life Writing: Essays and Lectures*. London: Routledge.

D'Amore, Jonathan. 2012. *American Authorship and Autobiographical Narrative. Mailer, Wideman, Eggers*. New York: Palgrave Macmillan.

Danielewicz, Jane. 2018. *Contemporary American Memoirs in Action. How to Do Things with Memoir*. Cham: Palgrave Macmillan.
Davis, David. 2020. "Faulkner." *American Literary Scholarship* (2020): 143–56.
Davis, Rocío G., ed. 2013. *The Transnationalism of American Culture. Literature, Film, and Music*. New York: Routledge.
Dewey, John. 1981. "The Live Creature." *The Philosophy of John Dewey*, ed. John J. McDermott, 525–40. Chicago: University of Chicago Press.
Dinan, Timothy G., Roman M. Stilling, Catherine Stanton, and John F. Cryan. "Collective Unconscious: How Gut Microbes Shape Human Behavior." *Journal of Psychiatric Research* 63 (April 2015): 1–9.
Dix, Hywel. 2018. *Autofiction and Cultural Memory*, London: Routledge.
Dobson, Kit. 2009. *Transnational Canadas: Anglo-Canadian Literature and Globalization*. Waterloo: Wilfrid Laurier University Press.
Douglas, Kate, and Anna Poletti. 2016. *Life Narratives and Youth Culture. Representation, Agency and Participation*. London: Palgrave Macmillan.
Dünne, Jörg, and Christian Moser, eds. 2008. *Automedialität. Subjektkonstitution in Schrift, Bild und neuen Medien*. Leiden: Brill.
Eakin, Paul John. 1985. *Fictions in Autobiography*. Princeton: Princeton University Press.
Eakin, Paul John. ed. 1991. *American Autobiography. Retrospect and Prospect*. Madison: The University of Wisconsin Press.
Eakin, Paul John. 1999. *How Our Lives Become Stories: Making Selves*. Ithaca: Cornell University Press.
Effe, Alexandra, and Hannie Lawlor. 2022. *The Autofictional: Aproaches Affordances, Forms*. London: Palgrave Macmillan.
Egan, Susanna, and Gabriele Helms. 2004. "Life Writing." In *The Cambridge Companion to Canadian Literature*, ed. Eva-Marie Kröller, 216–40. Cambridge: Cambridge University Press.
Farr, James R., and Guido Ruggiero. 2022. "Introduction: Historicizing Life-Writing and Egodocuments in Early Modern Europe." In *Historicizing Life-Writing and Egodocuments in Early Modern Europe*, ed. James R. Farr and Guido Ruggiero, 1–16. Cham: Palgrave Macmillan.
Fisher Fishkin, Shelley. 2005. "Crossroads of Cultures: The Transnational Turn in American Studies: Presidential Address to the American Studies Association, November 12, 2004." *American Quarterly* 57, no. 1 (March 2005): 17–57.
Fluck, Winfried, Donald E. Pease, and John Carlos Rowe, eds. 2011. *Re-Framing the Transational Turn in American Studies*. Hanover: Dartmouth College Press.
Fludernik, Monika. 1966. *Towards a 'Natural' Narratology*. London: Routledge.
Gao, Wei et al. "Gut Microbiome and Brain Functional Connectivity in Infants- A Preliminary Study Focusing on the Amygdala." *Psychopharmacology* 236, no. 5 (May 2019): 1641–51.
Gilmore, Leigh. 2001. *The Limits of Autobiography. Trauma and Testimony*. Ithaca: Cornell University Press.
Glaveanu, Vlad Petre. 2018. "Perspectival Collectives Futures: Creativity and Imagination in Society." In *Imagining Collective Futures: Perspectives from Social, Cultural and Political Psychology*, ed. Constance de Saint-Laurent, Sandra Obradovic, and Kevin Carriere, 83–106. Basingstoke: Palgrave Macmillian.
Glaveanu, Vlad Petre et al. 2019. "Advancing Creativity Theory and Research: A Socio-Cultural Manifesto." *The Journal of Creative Behavior* 54, no 3 (September 2020): 741–45.
George, Rosemary Marangoly. 1996. *The Politics of Home: Postcolonial Relocations and Twentieth-Century Fiction*. Berkeley: University of California Press.

Graham, Maryemma, and Mercedes Lucero. 2020. “Life Writing.” In *The Routledge Handbook to the Culture and Media of the Americas*, ed. Wilfried Raussert, Giselle Liza Anatol, Sebastian Thies, Sarah Corona Berkin, and José Carlos Lozano, 121–35. London: Routledge.

Greenblatt, Stephen. 1980. *Renaissance Self-Fashioning: From More to Shakespeare*. Chicago: The University of Chicago Press.

Gudmundsdottir, Gunnthorunn. 2017. *Representations of Forgetting in Life Writing and Fiction*. London: Palgrave Macmillan.

Hart, Kathleen. 2004. *Revolution and Women's Autobiography in Nineteenth-Century France*. Amsterdam: Rodopi.

Hebel, Udo, ed. 2012. *Transnational American Studies*. Heidelberg: Winter.

Helms, Gabriele. 2005. “Reality TV Has Spoken: Auto/biography Matters.” In *Tracing the Autobiographical*, ed. Marlene Kadar, Linda Warley, Jeanne Perreault, and Susanna Egan, 43–63. Waterloo: Wilfrid Laurier University Press.

Hinrichsen, Lisa. 2015. *Possessing the Past: Trauma, Imagination, and Memory in Post-plantation Southern Literature*. Baton Rouge: Louisiana State University Press.

Hornung, Alfred, ed. 2007. *Intercultural America*. Heidelberg: Winter.

Hornung, Alfred, ed. 2009a. “Transcultural Life-Writing.” In *The Cambridge History of Canadian Literature*, ed. Coral Ann Howells and Eva-Marie Kroeller, 536–55. Cambridge: Cambridge University Press.

Hornung, Alfred, ed. 2009b. *Autobiography and Mediation*. Heidelberg: Winter.

Hornung, Alfred, ed. 2013a. *American Lives*. Heidelberg: Winter.

Hornung, Alfred, ed. 2013b. “American Lives: Preface.” In *American Lives*, ed. Alfred Hornung, ix–xvii. Heidelberg: Winter.

Howells, W. D. “Editor's Easy Chair.” 1909. *Harper's Monthly Magazine;* (Cover Title: Harper's Magazine) 119, no. 713, October 1, 795. *ProQuest*. https://www.proquest.com/magazines/editors-easy-chair/docview/1301524746/se-2.

Jensen, Meg, and Margareta Jolly. 2014. *We Shall Bear Witness. Life Narratives and Human Rights*. Wisconsin: University of Wisconsin Press.

Jones, Elizabeth H. 2007. *Spaces of Belonging: Home, Culture, and Identity in 20th Century French Autobiography*. Amsterdam: Rodopi.

Kadar, Marlene, Susanna Egan, Linda Warley, and Jeanne Perreault. 2005. *Tracing the Autobiographical*. Waterloo: Wilfrid Laurier University Press.

Kaltmeier, Olaf, and Wilfried Raussert, eds. 2019. “Introduction.” In *Sonic Politics: Music and Social Movements in the Americas*, 1–18. New York: Routledge.

Kamboureli, Smaro, and Roy Miki. 2007. *Trans.Can.Lit. Resituating the Study of Canadian Literature*. Waterloo: Wilfrid Laurier University Press.

Kleinreesink, Esmeralda. 2017. *On Military Memoirs. A Quantitative Comparison of International Afghanistan War Autobiographies, 2001–2010*. Leiden: Brill.

Lanzendörfer, Tim. 2013. “Ecological Life Writing: E. O. Wilson's ‘Trailhead’ and the Collective Life.” In *Ecology and Life* Writing, ed. Alfred Hornung. Heidelberg: Winter. 287–98.

Latour, Bruno. 2005. *Reassembling the Social: An Introduction to Actor-Network-Theory*. New York: Oxford University Press.

Lenz, Günter H. 2011a. “Redefinitions of Citizenship and Revisions of Cosmopolitanism – Challenges of Transnational Perspectives.” *Symposium: Redefinitions of Citizenship and Revisions of Cosmopolitanism – Transnational Perspectives*, ed. Günter H. Lenz, Rob Kroes, Rüdiger Kunow, Alfred Hornung, and William Boelhower. *Journal of Transnational American Studies* 3, no. 1: 4–17.

Lenz, Günter H. 2011b. "Toward a Politics of American Transcultural Studies: Discourses of Diaspora and Cosmopolitanism." In *Re-Framing the Transnational Turn in American Studies*, ed. Winfried Fluck, Donald Pease, and John Carlos Rowe, 391–425. Hanover: Dartmouth College Press.

Maguire, Emma. 2018. *Girls, Autobiography, Media Gender and Self-Mediation in Digital Economies*. London: Palgrave Macmillan.

McLennan, Rachel. 2013. *American Autobiography*. Edinburgh: Edinburgh University Press.

Morrissette, Noelle. 2017. "Introduction: Biography of an Author, Biography of a Text: James Weldon Johnson's Ultimate American Work." In *New Perspectives on James Weldon Johnson's "The Autobiography of an Ex-Colored Man"*, ed. Noelle Morrissette, 1–20. Athens: University of Georgia Press.

Novak, Julia, and Caitríona Ní Dhúill. 2022. *Imagining Gender in Biographical Fiction*. Cham: Palgrave Macmillan.

Novak, Julia. 2017. "Experiments in Life-Writing: Introduction." In *Experiments in Life Writing: Intersections of Auto/Biography and Fiction*, ed. Lucia Boldrini and Julia Novak, 1–36. London: Palgrave Macmillan.

Nünning, Ansgar. 2022. "Fiktionalität, Faktizität, Metafiktion." In *Handbuch Biografie. Methoden, Traditionen*, Theorien, ed. Christian Klein, 29–35. 2nd ed. Berlin: J.B. Metzler.

Olney, James. 1980. *Autobiography. Essays Theoretical and Critical*. Princeton: Princeton University Press.

Ortiz-Vilarelle, Lisa. 2021. *Américanas, Autocracy, and Autobiographical Innovation. Overwriting the Dictator*. New York and Oxon: Routledge.

Poletti, Anna. 2020. *Stories of the Self. Life Writing after the Book*. New York: New York University Press.

Rak, Julie. 2009. Dialogue with the Future: Philippe Lejeune's Method and Theory of Diary. In *On Diary*, ed. Jeremy D. Popkin and Julie Rak, 16–26. Honolulu: University of Hawaii Press.

Rak, Julie. 2013. *Boom! Manufacturing Memoir for the Popular Market*. Waterloo: Wilfrid Laurier University Press.

Raussert, Wilfried. 2014. "Mobilizing 'America/América': Toward Entangled Americas and a Blueprint for Inter-American 'Area Studies.'" *fiar. forum for inter-american research* 7, no. 3: n. pag.

Raussert, Wilfried, and Reinhard Isensee. 2009. *Transcultural Visions of Identities in Images and Texts*. Heidelberg: Winter.

Renov, Michael. 2004. *The Subject of Documentary*. Minneapolis: The University of Minnesota Press.

Sayre, Robert F. 1978. "Autobiography and the Making of America." *The Iowa Review* 9, no. 2 (Spring 1978): 1–19.

Schaffer, Kay, and Sidonie Smith. 2004. *Human Rights and Narrated Lives: The Ethics of Recognition*. New York: Palgrave Macillan.

Sender, Ron, Shai Fuchs, and Ron Milo. 2015. "Revised Estimates for the Number of Human and Bacteria Cells in the Body." *PLoS Biol* 14, no. 8 (19 August): 1–14.

Shaffi, Sarah. 2022. "Have We Fallen Out of Love With Celebrity Memoirs?" *The* Guardian, December 9. https://www.theguardian.com/books/2022/dec/09/have-we-fallen-out-of-love-with-celebrity-memoirs.

Skonieczna-Żydecka, Karolina, Wojciech Marlicz, Agata Misera, Anastasios Koulaouzidis, and Igor Loniewski. 2018. "Microbiome – The Missing Link in the Gut-Brain Axis: Focus on Its Role in Gastrointestinal and Mental Health." *Journal of Clinical Medicine* 7, no. 521 (7 December): 1–18.

Smith, Sidonie, and Julia Watson. 1992. *De/Colonizing the Subject: The Politics of Gender in Women's Autobiography*. Minneapolis: University of Minnesota Press.

Smith, Sidonie, and Julia Watson. 2001. *Getting a Life: Everyday Uses of Autobiography*. Minneapolis: The University of Minnesota Press.

Smith, Sidonie, and Julia Watson. 2006. *Before They Could Vote. American Womens Autobiographical Writing, 1819–1919*. Wisconsin: The University of Wisconsin Press.
Smith, Sidonie, and Julia Watson. 2010. *Reading Autobiography: A Guide for Interpreting Life Narratives*. Minneapolis: The University of Minnesota Press.
Stitt, Jocelyn Fenton. 2017. "Postcolonial Life-Writing." In *The Bloomsbury Introduction to Postcolonial Writing: New Contexts, New Narratives, New Debates*, ed. Jenni Ramone, 177–91. London: Bloomsbury.
Thiemann, Anna. 2019. "Postmodernity." In *Handbook of Autobiography/Autofiction*, ed. Martina Wagner-Egelhaff, 778–806. Berlin and Boston: De Gruyter.
Thies, Sebastian, and Josef Raab, eds. 2009. *E Pluribus Unum? National and Transnational Identities in the Americas. Identidades nacionales y transnacionales en las Américas*. Berlin: LIT.
Twain, Mark. 2010. *Autobiography of Mark Twain, Volume 1. The Complete and Authoritative Edition*, ed. Harriet E. Smith, Benjamin Griffin, Victor Fischer, Michael Barry Frank, Sharon K. Goetz, Leslie Diane Myrick, and Robert Hirst. Oakland: University of California Press.
Vice, Sue. 2014. *Textual Deceptions: False Memoirs and Literary Hoaxes in the Contemporary Era*. Edinburgh: Edinburgh University Press.
Wagner-Egelhaaf, Martina, ed. 2019. "Introduction Autobiography/Autofiction Across Disciplines." In *Handbook of Autobiography/Autofiction*, 1–7. Berlin and Boston: de Gruyter.
Wallace, David Foster. 2006. "How Tracy Austin Broke My Heart." In *Consider the Lobster and Other Essays*, ed. David Foster Wallace, 141–55. New York; Little, Brown and Co.

Part I: **Life Narratives**

Alfred Hornung

Audre Lorde and Jack Kerouac in Mexico and the Creation of Transnational Life Writing

In this essay I proceed from two theoretical concepts and their application to life writing across the Americas, more specifically between the United States of America and Mexico. The first underlies Paul John Eakin's life-long autobiography scholarship and his assumption that the body is the basis of narrative and the self, i.e., the biological quality of life writing (Eakin 2020). The second is the recently advanced idea of the ontological turn of life writing, based on Bruno Latour's resistance to the division between the social/cultural and the natural in modern times who argues for their interrelation as hybrids from a non-modern perspective and advances a pluralist worldview (Latour 1993). The combination of these approaches lends itself to a discussion of life writings which originate from experiences by Americans in Mexico who were drawn to the land south of the border in pursuit of a more liberal and independent life. The attractiveness of Mexico is part of the culture of modernism when leftist Europeans and Americans sought the freedom of personal and artistic expression profiting from the revolutionary spirit and nature experience in Mexico. The group of authors and artists includes the French poet and critic Antonin Artaud, whose concept of the "theatre of cruelty" was inspired by his drug-induced experiences of the peyote ritual in the Sierra Tarahumara during his 1936 trip to Mexico, captured in his *The Peyote Dance* (Artaud 1948). The Russian avant-garde filmmaker Sergej M. Einstein traveled in Mexico in the 1920s to collect material for a major film project *¡Que viva México!* In the film begun in 1930, Eisenstein wanted to show Mexican culture from pre-Columbian civilization to the Mexican Revolution but could not complete it for lack of funds from the American writer and producer Upton Sinclair. Many of the modern American writers, artists, and critics were inspired by the leftist political climate and the multifarious history of Mexico, such as John Dos Passos who discovered the Mexican mural art during his 1926 visit and used it for his own paintings (Pizer, Nanney, and Layman 2016), or John Steinbeck who was particularly interested in the Mexican Revolution, visited Mexico frequently with a longer stay in 1935. Many Mexicans figure in his work, his novella *Pearl* is set in Mexico, and he provided the screenplay of *Viva Zapata!* (1952). Last but not least, Saul Bellow has his Jewish protagonist travel to Mexico in *The Adventures of Augie March* (1953) where he undergoes an essential training in his education and learns how to cope with vital forces of nature and love. All of these examples un-

https://doi.org/10.1515/9783111551968-002

derscore the political and cultural importance of Mexico for European avant-gardists and socially critical modern American writers who received innovative impulses for their works by personally experienced intercultural encounters. All of them prefigure Audre Lorde's and Jack Kerouac's Mexican experiences as a new way of embodied existence, an invigoration of their craft which instilled a new form of creativity for the practice of transnational life writing. I am applying the transnational idea, an established category of analysis in sociology, history, political science and cultural criticism, to autobiographical works whose writers transcend the political and cultural origin of their lives and envision a transcultural form of existence for themselves and their readers. In this sense, Lorde and Kerouac serve as exemplary cases of the field of transnational American studies (Morgan, Hornung, and Takayuki 2018).

For my analysis, I would like to examine the importance of the Mexican context in the 1950s for the creative evolution of Audre Lorde, the self-termed "black lesbian feminist mother lover poet" (Lorde 1997, 24) in relation to her parents' Carribean background and connect it with the Mexican experiences evident in the life narratives of the Canadian American beat poet Jack Kerouac. I will argue that both use the experience and writing of their Mexican encounters to transform the relationship between peoples and their worlds for the Americas and to advance social and political change. Both form part of a movement away from the elitist conception of an abstract modernism and the separation of nature and culture which accompanies the modern age. Instead, they opt for and advocate a communitarian lifestyle and recognition of diversity, which manifest themselves in the youth culture of the 1960s and the emergence of the postmodern resurrection of the body and a new body consciousness (Brown 1966). Their practice of life writing differs decisively from the conventional genre of autobiography as the recollection of chronological stages of a life related as a mnemonic activity from the perspective of old age. Their life narratives originate from their bodily experiences and connect biological nature and cultural activities. It is no surprise that the eminent autobiography critic Paul John Eakin shifts his scholarship from the conventional disciplinary boundaries of literary interpretations to the bodily quality of life writing after his encounter with António Damásio's neurobiological approach in the 1990s. Significantly, neuroscientist Damásio had seen the Cartesian split between the mind and the body as "Descartes' error" early on in his career (Damásio 1994) and later stressed the combination of mind and the autobiographical self (2010, 210). Damásio's idea that neural networks are responsible for the production of an autobiographical narrative leads Eakin to follow that "narrative is biological before it is linguistic and literary" (Eakin 2020, 32–34). This crucial role of narrative as bodily experiential quality is also the seedbed for the self and referential element of life writing. Audre Lorde's and Jack Kerouac's Mexican experiences in the 1950s represent prefigurations of their practice of life writing which anticipate the transformation of the genre in the 1980s when discourses of race, class, and gender motivate underprivileged and underrepresented people to write their lives in an effort of self-manifestation and emancipation. It is the impact of these new forms of experiential life writings which transform conventional autobiography scholarship into the inclusive genre of life writing. The unconventional lives that both writers live in Mexico determine the innovative interaction of the range of topics and their representations.

Audre Lorde's stay in Mexico in 1953–1954 is the topic covered in three chapters (20–22) of her autobiography *Zami: A New Spelling of My Name* (1982). This stay of the nineteen-year-old woman is motivated by personal, political, and literary reasons, which emerge from her life narrative of her childhood and adolescence. After having completed the strict Catholic education at primary and secondary schools, aggravated by the fact of being considered legally blind and hence being excluded from her pals in separate classes, she rebels first against her teachers and eventually against her parents, especially her mother, which leads to a separation from home. First sexual experiences with men and women make her realize her preference for women, regardless of color. The political situation determined by the intensifying persecution of leftish intellectuals in the McCarthy period and the corollary racial discrimination in the U.S. make her look for an alternative in the liberal climate of Mexico by intending to spend a year abroad at the University of Mexico. Her desire to go to Mexico intensifies after the death of her father and the alienation from her mother. Without concretely referencing the Mexican visits of any European or American author, the fate of Ethel and Julius Rosenberg becomes the final motivation to seek an alternative to the political system of the United States as her precursors. Mexico provides her with an unprecedented experience represented by the Mexican people, the Mexican landscape, and an apparently concomitant body feeling.

Her fascination with Mexico began with images of this Latin American country in geography books in high school and the impressions of a different landscape inviting a different lifestyle.

> I don't know why I was seized with such a desire to go to Mexico. Ever since I could remember Mexico had been the accessible land of color and fantasy and delight, full of sun, music and song. And from civics and geography in grade school, I knew it was attached to where I lived, and that intrigued me. (Lorde 1982, 147)

Over and above the pleasant features of the land, climate, and music-inspired life spirit, Lorde recognizes the different social climate, not mentioned in her high school textbooks. Walking the streets of Mexico City, she encounters the "brown faces" which have "a profound and exhilarating effect" upon her and she blends in because she is "noticed and accepted without being known," a feeling of indiscriminate acceptance which gives her "a social contour" (Lorde 1982, 154). She quickly moves from Mexico City to Cuernavaca "a haven for political and spiritual refugees from the north" (158), victims of McCarthyist purges, such as the red-baited Hollywood Ten, leftist writers, and artists. The simultaneous presence of FBI agents in search of allegedly complicit collaborators hiding in Mexico, reminds her of her own engagement in preventing the unjust execution of Ethel and Julius Rosenberg shortly before her departure for Mexico on 19 June 1953. For Audre Lorde, this new life of American expats in Cuernavaca is exemplarily embodied by Eudora whom she quickly knows to be lesbian and calls "the most fascinating woman I had ever met" (162).

Previous sexual experiences with white and colored women in the greater area of New York, which she relates to her new acquaintance, are surpassed by the relationship with the 48-year-old Eudora whose unusual biography seems to heighten the intensity of their emotions. Eudora is the youngest child of an oil-worker family in Texas, who had suffered from Polio as a child, fought to be admitted to the University of Texas at Austin as the first woman in 1925, and had lived in a tent with her rifle and a dog because there was no housing for women on campus. Pursuing a career as a journalist, she followed her lover Franz to Chicago and decided to move to Mexico when he left her to marry a foreign correspondent in Istanbul (Lorde 1982, 163). In Cuernavaca, she started a bookstore with her new lover Karen, which became a "rallying place for disaffected americans" (163). Audre records her conversations with Eudora about literature, about Eudora's translations of texts on the history and ethnology of Mexico, about her favorite poet, Walt Whitman, and her plan to conduct research on African and Asian influences in Mexican art. These conversations become a prelude to their passionate relationship. In great detail, Lorde describes the new bodily and spiritual feelings of their intimate encounters and resultant feeling of being loved:

> It was in Mexico that I stopped feeling invisible. In the streets, in the buses, in the markets, in the Plaza, in the particular attention within Eudora's eyes. Sometimes, half-smiling, she would scan my face without speaking. It made me feel like she was the first person who had ever looked at me, ever seen who I was. And not only did she see me, she loved me, thought me beautiful. This was no accidental collision. (173)

In many respects, Eudora offers to Audre Lorde the central lessons of life which the young student had not learned in her classroom education nor experienced in her previous American love life. The scope of Eudora's program of learning by doing is all-inclusive and combines mind and body knowledge which guarantees complete satisfaction. It includes visits to the locations of the alternate scene, the meeting of free spirits and political minds, the recommendations for excursions and new practices, the introduction to new restaurants, bars, and eating preferences, at times culminating in excessive drinking and drug habits.

Audre Lorde's fascination with the different natural and social environment of Mexico is shared and simultaneously lived by the Beat Generation, especially the Franco-Canadian American writer Jack Kerouac, one of its major representatives (cf. Calonne 2022, 115–36). Unbeknown to each other, they reject the consensus-based American system and believe in the emancipatory potential of alternate forms of life and art in the 1950s. The drug-induced trip of Beatniks from New York to San Francisco, profusely described in the major text of the avant-garde movement, Jack Kerouac's *On the Road* (1957), represents the vison of a shared communal life resulting in the free love of the hippie movement. Kerouac's first trip to Mexico in a car with Neal Cassady in 1950 is followed by several other trips throughout the decade. These journeys are the logical conclusion of the Beat Poets' search for an authentic form of life, which transcends racial, religious, and social boundaries and suggests – in Bruno Latour's words – new negotiations of peoples and worlds. The beginning public awareness of Mexican people in the Southwest, especially in California, and the embracement of cultural differences, the public display of the intercultural application of the political impact of murals as part of the Mexican Revolution coalesced with the emergence of the Chicano movement in the pursuit of transforming American society. The message of Allen Ginsberg's poem "Howl" and its reading in the City Lights bookstore in San Francisco in the fall of 1955 represented the spirit and lifestyle of the assembled Beatniks which were legally fought in the U.S., but possible ways of life in Mexico. Kerouac's book-length experimental poem *Mexico City Blues* originates in such a free Mexican environment. It represents a form of life writing which in its thematic and formal aspects recalls the combination of body and soul and all-inclusiveness, as best expressed by the Beatnik's most celebrated poet, in Walt Whitman's autobiographical *Song of Myself* (cf. Hornung 2019). Kerouac and the Beat Generation embrace Whitman's multimedia and diversity panorama and extend the geographical and cultural parameters to a transnational context. Throughout the 242-page poem, Kerouac covers central political topics from the 1930s to the 1950s: the Depression, the red decade, and FDR's social programs for starving people, the first nuclear explosions in Nevada, the emergence of McCarthyism perpetrated by the Republican party, the battle of political systems between Communism, Capitalism, and Democracy. From the perspective of his Mexican location, Kerouac critiques American patriotism, feels homeless (Kerouac 1990, 216), and impedes Joe McCarthy to "remove [his] my name from the list, and Buddha's too" (125). The universal claim of a Buddha Consciousness is based on the equality of all religions, extends to the biosphere, the "love of all sentient beings" and the bodily lived practice of the Christian religion, which ranges from the life of Jesus Christ "born in a barn because the inn was full. Egyptian, /Babylonian, /African" (68) to St. Francis's caring for birds (232). This also includes Kerouac's association of Christ with an Eastern context and Eastern philosophies, opening up a multiethnic alliance of all humankind (70). It also connects with his own biographical background as a Canuck in Massachusetts and the French connection, which he explored in Brittany, as well as the Catholic education received from his mother after his father's premature death. In a form of cultural appropriation, he calls himself "only an Apache" (149) and eventually claims to abandon all color, creed, or social differences: "Blue, Black, race, grace, face, I love ye" (197). As in the case of Audre Lorde, alcohol and drugs enable Kerouac to propagate a "universal mind" (183) and to opt for existence over essence which for him is "like absence of reality" (105). This denial of reality he also sees in the ready American acceptance of Sigmund Freud's psychoanalysis as a Gospel which relegates all forms of queerness to the "Chart of Mad talk" (175). It is not surprising that Kerouac ends his poem with a eulogy of the deceased Blues musician Charlie Parker (239–42) whom he calls "the perfect musician" (239), who is "Musically as important as Beethoven / Yet not regarded as such at all" (240). This embracement of Black music and its equality with the so-called standard of white classical music have critics led to suggest:

> The blues, if it did not make Jack black, it did give him color; it gave identity to a taciturn, sensual, and mystical eighteen-year-old whose commitment to the art was as serious as it was nebulous in its definition: the sound, the beat and rhythm of the voice and the human experience were to haunt him forever. (Ortiz Garay 2021)

Both Kerouac and Lorde mention the appeal of Black music for other singers as well as the music industry. Kerouac refers to the transformation of Al Jolson, the son of a Jewish rabbi, who had emigrated from Lithuania and had prepared his son, a gifted singer in the synagogue, to follow his position in the Jewish community. Instead, Al becomes "The Great Jazz Singer" and is the star in the 1927 movie *The Jazz Singer* in which he performs Black music as a Jewish cantor (Kerouac 1990, 116). In a critical assessment of the appropriation of Black culture by "straight white america" in the 1950s, Audre Lorde records: "For some, Elvis Presley and his stolen Black rhythms became arch-symbols of the antichrist" (Lorde 1982, 220). Last but not least and retroactively, Kerouac relates the Indigenous Mexican population to Black music and speaks of Aztec Blues, which he equates in one line with an "Imitation of Pound" and the experimental Cantos of the expatriate American poet (Kerouac 1990, 207). The recognition of the innovative potential of Black music by the mainstream population and its commercialization by the white music industry have become major topics and concerns of Transnational American Studies scholarship (cf. Raussert 2000, Raussert et al. 2020).

Audre Lorde's and Jack Kerouac's avant-gardist contribution to the creation of a new form of life writing which anticipated the transformation of conventional autobiographical texts into the bodily based and diverse presentations of selves in the 1980s reflects the theoretical assumptions of both Bruno Latour and Paul John Eakin. In this sense, both Lorde and Kerouac practice an ontological form of the representation of lived experiences in which they combine the relationships of peoples and their worlds hemispherically and transnationally with the implicit intention to create an awareness of and new sensibility for alternate forms of life and to advance the social and political situation in the U.S. and elsewhere. Based on their Mexican experiences, they surpass the forms of life writing assembled by Ricia Anne Chansky's collection of essays *Auto/Biography across the Americas: Transnational Themes in Life Writing* (2017), which constituted the foundation of the regional chapter of the International Auto/Biography Association, IABA-the Americas. Lorde's and Kerouac's autobiographical texts also transcend the division of life writings from Latin American and Indigenous authors from those of North American writers in contact with Latin America, the topic of Maryemma Graham and Mercedes Lucero's article on "Life Writing" in *The Routledge Handbook to Culture and Media of the Americas* (Graham and Lucero 2020). Although Audre Lorde's *Zami: A New Spelling of My Name* appears to be a conventional autobiography in which the writer narrates her life from childhood to adolescence with a focus on the Mexican stay and the changed embodied consciousness when she returns to New York City, I would like to claim that it represents a new approach already indicated in the title in the change of her "slave name" to a Caribbean one taken from the Carriacou women in her mother's Grenada where "women love each other" in the absence of their sea-faring men (Lorde 1982, 255). Both, the issue of lesbianism and the relation of being gay and black are the central focus of her autobiography and the new identity she brings back to and lives in New York (180). It is not the hidden and marginal life of the subaltern (224–26), but the self-assured existence of an openly gay and transnational citizen (cf. Bolaki and Broeck 2015). Next to this racial and sexual advance in the seemingly conventional form of her autobiography is a crucial moment in her erotic intimacy with Eudora in Cuernavaca. Eudora had mentioned that her operation for breast cancer led to the separation from her bookstore partner Karen. In their first love making, the sight and touch of this "scarred chest" becomes the initiation to their intimate relation:

> The pale keloids of radiation burn lay in the hollow under her shoulder and arm down across her ribs. I raised my eyes and found hers again, speaking a tenderness my mouth had no words yet for. She took my hand and placed it there, squarely, lightly, upon her chest. Our hands fell. I bent and kissed her softly upon the scar where our hands had rested. (167)

The combination of medical language and emotional bonding centering on the scar is revealing and permits – in my analysis – the reinterpretation of Lorde's autobiography as a link to her life narratives in poems and prose from the 1950s on and a demonstration of what Mita Banerjee calls "the power of autobiography" in *Medical Humanities in American Studies: Life Writing, Narrative Medicine, and the Power of Autobiography* (2018). Two years before the publication of *Zami*, Audre Lorde had published her *Cancer Journals* (1980) in which she relates her operation for breast cancer in an effort to account for the pain and routine medical treatment in "The very bland whiteness of the hospital" (46) and her life thereafter (cf. Hornung 2020, 2024). On the one hand she wants to connect in solidarity with other women with a similar medical experience, on the other hand she relates the disease of cancer to that of racism: "racism/cancer: in both cases to win the aggressor must conquer, but the resisters need only to survive. How do I define my survival and on whose terms?" (Lorde 2012, 109:46). To overcome both, a process of healing is necessary which begins with the writing of the seemingly conventional *Zami* with its focus on Eudora and the invention of a new genre name: "biomythography." Recalling the stages of her life with the crucial encounter of Eudora in Cuernavaca represents another way in which to cope with her own breast cancer. Similarly, she sees her stay as a guest professor at the John F. Kennedy Institute in 1984 with her partner Gloria I. Joseph only two years later as a continuation of her battle with "racism/cancer." Giving the so-called "war babies" in her class a new name "Afro-German" and initiating their life writing (Oguntoyee, Opitz, and Schulz 1992) is part of a transnational engagement, advocating the inclusion of all Black people in her scheme of a new world.

> . . . our connections [to Africa] are real. In addition to shaping our individual national identities within the diaspora, the question pertains for African Americans and African Europeans and African Asians alike: what is our relationship to Africa as a whole? What should be our input into and expectation from strong and independent African states? What is our role as nationals in the liberation struggle of southern Africa? What is our responsibility? (Lorde 2009, 172–73)

It is no surprise that shortly before her death in the American Virgin Island Saint Croix in 1992, Audre Lorde underwent an African ceremony in which she adopted the name of *Gambda Adisa*, "She Who Makes Her Meaning Known."

In his *Mexico City Blues*, Jack Kerouac attempts to realize an embodied and ontological existence of an interconnected transnational community. The form he chose for his long poem expresses these intentions. It consists of 242 Choruses, one page for each chorus. Each of them presents a medley of ideas without any specific rhyme or meter. Instead, Kerouac follows the improvisational and rhapsodic pattern of the blues and jazz. It is the application of his ideas which he had expressed in his essays "Essentials of Spontaneous Prose" (1959) and "Belief and Technique for Modern Prose" (1995) in which he announced his writing program as a move away from chronological or rational sequences or the conventional grammar of language in favor of the free flow of his embodied imagination. Rightly, the Beat poet Allen Ginsberg called *Mexico City Blues* "A spontaneous bop prosody and original classic literature," quoted on the cover of the volume, which in my interpretation constitutes an original form of transnational life writing.

The relation to musical arrangements also shows in Kerouac's choice of "chorus" for each entry, a term which implies the repetition of certain sections by a communal consonance of singers including all ethnicities, East and West, North and South. To a certain extent, these choruses call up Ezra Pound and his elaborate Cantos with a universal range and covering thousands of years of Chinese dynasties on several pages. At the same time, Kerouac goes beyond the single voice of Pound's canto – "I sing" – in the simple side-by-side position and multiethnic correlation of "Aztec Blues – Imitation of Pound" (Kerouac 1990, 207). Along with the creation of onomatopoetic words, minor prefigurations of graphic narratives, *Mexico City Blues* represents an all-inclusive autobiography which concludes – in classical autobiography style – with the intimation of death and a consolation for the future. His final eulogy for Charlie Parker in four choruses exemplifies both his universal topic and musical technique of life writing. The 239th Chorus begins with a line "Charley Parker Looked like Buddha" and Parker's favorite expression, "All is Well," which echoes the peaceful Buddha spirit. Chorus 240 opens with the reference to Beethoven: "Musically as important as Beethoven." Chorus 241 determines the universal reach of the three related figures:

> Not to be measured from here
> To up, down, east, or west –
> – Charley Parker, lay the bane,
> off me, and every body.

In line with the spirit of the Beat movement, the image of death receives a positive note, when the speaker of the poem, the autobiographical self of Jack Kerouac takes on the role of "a guard" for a transformed humankind.

> Stop the murder and the suicide!
> All's well!
> I am the Guard. (Chorus 242)

For both Audre Lorde and Jack Kerouac, their Mexican experiences become the platform for the unexpected but hoped-for transformation of their lives and for the formation of transnational life writings. Their desire to go to Mexico in the

1950s is the search for freedom from the social and political conditions in the United States which continues to adhere to a consensus model of the majority white population and to discriminate against people of color and of different political and personal persuasions. In order to establish and realize their alternate forms of life they long for a different cultural space and natural environment, which a score of intellectuals and artists from Europe and the U.S. had found in Mexico before them throughout the first half of the twentieth century. Living in a non-English language country, whose European colonial history pales against the presence of the impressive number of the Indigenous population and its age-old origin visible in so many impressive pyramids, myths, and rituals provides a multi-ethnic and divers background which allows them to develop their own emancipatory profile in a shared community. It is this extraterritorial space in which the national affiliation of the American citizens recedes and is replaced by a transnational alliance of underprivileged and underrepresented people. In a double sense, this is the fertile ground on which alternate lifestyles thrive and lead to innovative forms of creativity and transnational life writing. It extends the avant-gardist enrichment of Antonin Artaud, Sergej Eisenstein, and modern American authors found in Indigenous traditions and liberal political views in Mexico to a field of writing which incorporates personal experiences and enacts the vision of borderless transnational communities.

Works cited

Artaud, Antonin. 1976. *The Peyote Dance*. 1948. Transl. Helen Weaver. New York: Farrar, Straus and Giroux.

Banerjee, Mita. 2018. *Medical Humanities in American Studies: Life Writing, Narrative Medicine, and the Power of Autobiography*. Heidelberg: Universitätsverlag Winter.

Bellow, Saul. 1953. *The Adventures of Augie March*. New York: Viking Press.

Bolaki, Stella, and Sabine Broeck, eds. 2015. *Audre Lorde's Transnational Legacies*. Amherst: University of Massachusetts Press.

Brown, Norman O. 1966. *Life Against Death: The Psychoanalytic Meaning of History*. Middletown, CT: Wesleyan University Press.

Chansky, Ricia Anne, ed. 2017. *Auto/Biography Across the Americas: Transnational Themes in Life Writing*. New York: Routledge.

Calonne, David Stephen. 2022. *The Beats in Mexico*. New Brunswick, NJ: Rutgers University Press.

Damásio, António. 1994. *Descartes' Error: Emotion, Reason, and the Human Brain*. New York: G. P. Putnam.

Damásio, António. 2010. *Self Comes to Mind: Constructing the Conscious Brain*. New York: Pantheon Books.

Eakin, Paul John. 2020. *Writing Life Writing: Narrative, History, Autobiography*. New York: Routledge.

Eisenstein, Sergej M. 1930. *¡Que viva México!* Unfinished.

Graham, Maryemma, and Mercedes Lucero. 2020. "Life Writing." In *The Routledge Handbook to Culture and Media of the Americas*, ed. Wilfried Raussert, Giselle Liza Anatol, Sebastian Thies, and Sarah Corona Berkin, 121–35. New York: Routledge.

Hornung, Alfred. 2019. "Life Writing and Diversity: Walt Whitman's Song of Myself." In *Revisiting Walt Whitman on the Occasion of his 200th Birthday*, ed. Winfried Herget, 111–20. Berlin: Peter Lang.

Hornung, Alfred. 2020. "Zhuanji Zhishi Yu Xushi Yixue: Chuangzao Kuayue Guojie De Ziwo." [Life Writing Knowledge and Narrative Medicine: Creating the Transnational Self]. *Journal of the Graduate School of Chinese Academy of Social Sciences* 4 (July): 81–90.

Hornung, Alfred. 2024. "Travel, Disease and Life Writing as World Literature: Jack London, Audre Lorde, Gao Xingjian." In *Life Writing as World Literature*, ed. Helga Lenart-Cheng and Ioana Luca. New York: Bloomsbury Publishing.

Kerouac, Jack. 1990. *Mexico City Blues*. New York: Grove Press.

Kerouac, Jack. 1959; 1995. "Essentials of Spontaneous Prose" and "Belief and Technique for Modern Prose." In *The Portable Beat Reader*, ed. Ann Charters, 57–58; 58–59. New York: Viking.

Latour, Bruno. 1993. *We Have Never Been Modern*. Transl. Catherine Porter. Cambridge, MA: Harvard University Press.

Lorde, Audre. 1982. *Zami: A New Spelling of My Name*. Freedom, CA: The Crossing Press.

Lorde, Audre. 1997. *The Cancer Journals*. Special Edition. San Francisco: aunt lute books.

Lorde, Audre. 2009. *I Am Your Sister: Collected and Unpublished Writings of Audre Lorde*, ed. Rudolph P. Byrd, Johnnetta Betsch Cole, and Beverly Guy-Sheftall. New York: Oxford University Press.

Lorde, Audre. 2012. *The Berlin Years, 1984–1992*. Dir. Dagmar Schultz. Germany.

Morgan, Nina, Alfred Hornung, and Tatsumi Takayuki, eds. 2018. *The Routledge Journal of Transnational American Studies*. New York: Routledge. Paperback 2022.

Oguntoye, Katharina, May Opitz, and Dagmar Schultz, eds. 1992. *Farbe Bekennen: Afro-deutsche Frauen auf den Spuren ihrer Geschichte*. 1986. Fischer: Frankfurt am Main.

Opitz, May, Katharina Oguntoye, and Dagmar Schultz, eds. 1992. *Showing Our Colors: Afro-German Women Speak Out*. Trans. Anne V. Adams. Amherst: University of Massachusetts Press.

Ortiz Garay, Andrés. 2021. "The Story of Writer Jack Kerouac and his Relationship with Mexico." *Mexicanist*. August 19. https://www.mexicanist.com/l/american-novelist-jack-kerouac/.

Pizer, Donald, Lisa Nanney, and Richard Layman. 2016. "Preface to Section 7 Mexico 1926–32." In *The Paintings and Drawings of John Dos Passos: A Collection and Study*. Liverpool: Liverpool University Press, 40–41.

Raussert, Wilfried. 2000. *Negotiating Temporal Differences: Blues, Jazz and Narrativity in African American Culture*. Heidelberg: Universitätsverlag Winter.

Raussert, Wilfried, Giselle Liza Anatol, Sebastian Thies, and Sarah Corona Berkin, eds. 2020. *The Routledge Handbook to Culture and Media of the Americas*. New York: Routledge.

Raussert, Wilfried, and Giselle Liza Anatol. 2020. "Introduction – Literature and Music of the Americas." In *The Routledge Handbook to Culture and Media of the* Americas, 17–22. New York: Routledge.

Steinbeck, John. 1952. Screenplay *Viva Zapata!*

Whitman, Walt. 1959. "Song of Myself." In *Complete Poetry and Selected Prose*, ed. James E. Miller, Jr., 25–68. Boston: Houghton Mifflin Company.

Frank Mehring

Prismatic Life Writing: Drawing the Schwarzwald in Desierto de los Leones. The German American Artist, Designer, and Teacher Winold Reiss in Mexico

In a remarkable confluence in the year 2022, two new book publications and three exhibitions, one of them major, shine a new light on a largely forgotten treasure: the work of the German American artist, designer, and teacher Winold Reiss (1886–1953).[1] The first comprehensive scholarly anthology explores *The Multicultural Modernism of Winold Reiss*. An exhibition at the New-York Historical Society titled *The Art of Winold Reiss: An Immigrant Modernist* emphasizes the transatlantic dimension of Reiss in the title calling him an *Immigrant Modernist.* This pioneer of transatlantic and transnational modernism is a remarkably versatile German American painter, designer, and teacher, described in *Du Pont Magazine* (1931) as a "modern Cellini" who defies instant categorization and labels. Reiss remains a mystery to many art critics. What was true in the 1930s holds in the new millennium: "Winold Reiss will not be classified" (Sampson 1931, 4) What makes it so worthwhile to engage with this artist who defies easy categorization and seems to have fallen through the cracks of art historical scrutiny – though back in his day he was described as 'a modern Cellini'?

Reiss's modernist aesthetic of portrait drawing, liberated from the confines of nineteenth-century genre, points to a future more than half a century away, Jeffrey Stewart explains in his essay, referring to Reiss's portraits of African Americans in 1920s New York. Reiss's perspective on the realities of American confrontations of ethnicity and institutionalized racism informed his conviction that art can make an

1 *The Multicultural Modernism of Winold Reiss* (Mehring 2022) presents essays by art historians and cultural scientists and approaches Reiss's oeuvre from national and transnational perspectives. The book is based on the first international Winold Reiss Symposium organized by Frank Mehring in Berlin in 2011. *The Art of Winold Reiss: An Immigrant Modernist* (Kushner 2021) accompanied a major exhibition at the New-York Historical Society in New York City, on view from July 1 to October 9, 2022, with the main focus on Reiss's career in New York. "F. Winold Reiss: Portraits of the Blackfeet of Glacier Park" exhibits 24 Reiss works from the permanent collection of the C.M. Russell Museum in Great Falls, Montana and is on view there from May 26, 2022, until May 2023. "Connections: The Blackfeet and Winold Reiss" was on view at the Museum of the Plains Indian in Browning, Montana from May 27 to September 30, 2022.

https://doi.org/10.1515/9783111551968-003

important contribution to resolving the discrepancy between a democratic promise of freedom and a deplorable practice of "White" hubris and xenophobia.[2]

Born in Karlsruhe in 1886, Winold Reiss spent the formative period of his childhood and youth in the Black Forest region in and around Freiburg (1899–1910) after living in Leipzig (1888–1892) and Stuttgart (1892–1899). In October 1910, Reiss moved to Munich and studied with renowned artists Julius Diez (at the School of Applied Arts) and Franz von Stuck (at the Academy of Fine Arts.) Reiss emigrated to the United States in 1913 – about 9 months before the outbreak of World War I in Europe – and became an influential figure in transatlantic modernism in the 1920s and 1930s. Reiss founded his own art school and design studio in New York's Greenwich Village and taught at the New York School of Applied Design for Women, the Keramic Society and Design Guild of New York, and the School of Architecture at New York University. He conducted summer schools in Woodstock, New York, and later in Glacier Park, Montana, where he drew many portraits of Native Americans in their own cultural setting.

During his time, Reiss's work was widely exhibited and appeared in publications such as *Scribner's Magazine*, *Century Magazine*, *The Forum*, and *The Survey Graphic*; he also co-founded the art magazine *Modern Art Collector.* His pastel portraits of Harlem Renaissance artists and intellectuals were essential to the success of the anthology *The New Negro: An Interpretation* (Locke 1925b) and have become iconic. Reiss depicted his sitters – among them philosopher and civil rights activist W.E.B. Du Bois; the first internationally acclaimed African American concert artist Roland Hayes, and the anthropologist and author Zora Neale Hurston – not as distorted caricatures but as dignified, proud Americans, in a visual language that was visionary for his time and continues to resonate today.

Considering the wide array of professional activities in portraiture, graphic design, interior design, or mural painting in different parts of the world, the ques-

2 The scant recognition of Reiss's innovative approaches to poster art, lettering, and interior design stems to a certain degree from the continuing stigma of commercial art among art historians. Jeffrey C. Stewart was among the first to describe Winold Reiss's life and his dedication to drawing ethnic minorities in North America in the catalogue *To Color America: Portraits by Winold Reiss* (Stewart 1989), which accompanied a major exhibition curated by him at the Smithsonian Institution's National Portrait Gallery in 1989. See Jeffrey Stewart, ed., *To Color America: Portraits by Winold Reiss* (1989) with companion volume *Winold Reiss: An Illustrated Checklist of his Portraits* (1990). Among the comparatively small exhibitions of Reiss's artwork are *Winold Reiss, Native American Portraits*, exhibition catalogue with an essay by Scott J. Tanner and introduction by Richard V. West (2000); Marion Grant, *Diverse Reflections: Portraits by Winold Reiss* (1997); *Native Faces: Winold Reiss* (1997), exhibition catalogue with an essay by John Heminway; and Ray W. Steele, ed., *Winold Reiss, Portraits of the Races*, exhibition catalogue with text by Paul Raczka and introduction by Ned Jacob (1986).

tion arises: What options do we have then to describe his life and work? I suggest a new concept of life-writing that focusses on themes with alternative overlapping timelines.

As an academic concept, life writing refers to a genre that encompasses autobiographical or biographical narratives, memoirs, diaries, letters, and other forms of written accounts that focus on personal experiences and the exploration of one´s life. It involves the documentation and interpretation of individual or collective life stories, offering insights into the human condition, personal identities, and sociocultural contexts. It's essence lies in the "representation of an individual self," as Alfred Hornung claims (2001, 38).

As a biographical approach to Winold Reiss, I suggest a novel concept of life writing that pertains to the multifaceted and fragmented nature of autobiographical accounts. I suggest describing his life experiences through a diverse array of perspectives, reflecting the complexity and subjectivity of a human life. This concept of life writing incorporates various viewpoints, identities, and narratives, resembling the facets of a prism that refracts light into different colors. This will allow me to capture the nuances, contradictions, and diversity of Reiss's personal and artistic experiences on both sides of the Atlantic, challenging the singular, linear narrative often associated with more traditional, chronological autobiographical works.

I take my cue from the physical phenomenon of light manifesting in seven bright colors as it passes through an optical prism. I suggest a *prismatic life writing*. Rather than tracing a person's life chronologically, I propose to split up the narrative into thematic vignettes which, taken together, compose the biography. This (white) light is recomposed through overlaps of the different colors (in a metaphorical fashion) of the vignettes. The vignettes offer entrances to the biography without claiming to be the only or central gateway. At the center of this approach are stories (in the sense of Alfred Hornung) told by a human subject "on different occasions in a number of different forms" (Hornung 2013, x).

The optical prism metaphor in approaching the complexity of Winold Reiss's life and art is particularly appropriate as he used the very word "prismatic" to describe his interior design approach for the mural decorations in the so-called New Tea Room at the Restaurant Crillon, 15 East 48th Street, New York City (ca. 1923). He called this room which has references to his encounters with Indigenous Peoples and Mexican culture (Figure 1) the "Prismatic Room."[3] I will explore in more detail how Reiss framed these encounters in Mexico as a kind of transna-

3 Ford Peatross has recently offered an overview of Reiss's design work in his article "Multicultural and Modern Design: The Singular Vision of Winold Reiss, 1906 to 1952" (2022).

Figure 1: Winold Reiss, *Mural decorations, New Tea (Prismatic) Room, Restaurant Crillon*, 15 East 48th Street, New York City. Ca. 1923–1926. Preliminary study, elevation, graphite, and tempera on illustration board (detail). Library of Congress, Prints & Photographs Division, Winold Reiss Design Collection.

tional epiphany. I argue that his travels through Mexico in 1920 are central to understand this visual approach to themed spaces in public places. Some of his so-called "Imaginative Drawings" explore the potential of presenting the human body in refractions of light such as an untitled watercolor of a woman kissing a man, which supports the prismatic function of color as a key to unlock multi-perspectival approaches to a person's life (Figure 2).[4]

In my prismatic approach, I will be focusing on four "optical prism gateways" which then refract the light into different colors: locations, social networks, media, and art. These gateways will allow for a different form of life writing that

4 I am also inspired by the way Henry David Thoreau captured his time living on Walden Pond in an innovative fashion. *Walden. An Annotated Edition* (Thoreau 1995) can be understood as a form of life writing split into 13 chapters – or vignettes – based on specific gateways such as Economy, Where I lived, and What I lived for, Reading, Sounds, Solitude, Visitors, The Village, Baker Farm, Higher Laws, House Warming, Former Inhabitants, Winter Visitors, Winter Animals, Spring (followed by a conclusion). While loosely structured around different seasons, the chapters capture life at Walden Pond focusing on different fields such as economy, social networks, locations, or media.

Figure 2: Winold Reiss, Untitled, 24 x 19", watercolor on paper, n. d.

works with what Thomas Bender called a "matrix of entanglements" to illuminate how Reiss the artist, the designer, the teacher worked and how ideas expressed themselves in different socio-cultural and urban environments. Bender's warning that biographies of academics, e.g., often suffer "from the abundance of evidence on what they read and what they wrote," is also relevant for the biographer of Reiss (Bender 2013, 27). What is important is to provide different inroads to an artist's life through the contextualization of what I call gateways of creativity.

These gateways then show themselves refracted as follows:

1. Locations: Black Forest, Munich, the Atlantic, New York City, Browning Montana, Mexico, Hudson/NY
2. Social networks: family, wife, love affairs, friendships, students, business partners, artistic networks
3. Media: photography, books, magazines, diaries, letters, advertisements, exhibitions,
4. Art: design, drawings, murals, printing, teaching, writing, music.

Transcultural homesickness: conflating Mexico and the Black Forest

Reiss's manuscript *Mexico Diary* offers a glimpse into the artist's mind and into the shaping of what would soon be developed most forcefully in the visual narrative of *The New Negro: An Interpretation* (1925).[5] The diary is written in German and contains three elements: (1) matter-of-fact travel information regarding Mexican places, spaces, and people; (2) meditations on the function of color in Mexican landscapes (these passages also identify people and locations portrayed by Reiss); and (3) critical observations on borders and philosophical reflections on issues of gender constructions, racial conflicts, and the function of art in metropolitan cities.[6]

The trip to Mexico promised spiritual renewal. It led him from Mexico City to various locations such as Xochimilco, Guadalupe, Chapultepec, Puebla, Oaxaca, and Tepotzotlán. Earlier that year, after drawing the Blackfeet Indians on their reservation in Browning Montana, he had paid a brief visit to Taos, New Mexico. In his article "Draughtsmanship and Racial Types/Mexican Character Studies by Winold Reiss," Morris De Camp Crawford describes that the Taos experience "affected Reiss like a powerful narcotic" (1921, 29) and convinced the artist to focus his ambitions on the Southwest and Mexico.[7]

On his walk from La Venta to the former monastery of Desierto de los Leones on October 14 (cf. Figure 3, 4, and 5), Reiss came face to face with Mexican revolutionaries. In a striking conflation of memories of the Black Forest and the encounter with Mexican revolutionaries, Reiss began sketching a series of soldier drawings.

---

5 The following part builds on parts published in the trilingual edition of *The Mexico Diary* (Mehring 2016).

6 There exist three sets of notes Winold Reiss wrote about his Mexico trip, all in German with a sprinkling of English here and there. The most complete set ("Version I") covers his entire journey from October 5 to December 12, 1920. Written in longhand, with ink, pencil, and color pencil on various loose sheets – mostly lined 2-hole punched 9 1/2" x 7 1/2" (24 x 19 cm) note paper – and without title or page numbers, these notes appear to have been written on the spot, at times in haste. The second set ("Version II") is an edited version of the opening part of Version I, covering the days of October 5 through 9. It is written longhand in ink on loose sheets of 10 1/4" x 7 1/4" (26 x 18.4 cm) lined paper, without title or page numbers, and the year in the opening entry is misstated as "1922." The third set ("Version III") is handwritten in block letters in a bound 8 1/2" x 7" (20.3 x 16.5 cm) notebook, in ink with further edits in pencil, and features the title *Meine Reise durch Mexico*. This text is similar to Version II, and its second entry misstates the year as "1921." All quotations of the *Mexico Diary* refer to the English version edited and published by the author: *The Mexico Diary: Winold Reiss between Vogue Mexico and Harlem Renaissance* (Mehring 2016).

7 The article was first published in *Arts & Decorations* (May 1921, 9) and was reprinted in the catalogue for the exhibition at the Anderson Gallery.

Thursday, October 14, 1920. One hour on the electric to La Venta. From there by foot to the Desierto monastery. Fir forest like in the Black Forest. The same resinous air. Always uphill. I feel the altitude, have difficulty breathing now and then and cannot go as fast as usual. Once again real tall fir trees, Heimat [homeland], and all that in Mexico. Such contrasts! We reached the monastery in forty-five minutes. Entirely in ruins the monastery presents a rare and original picture amidst this high forest. We were up on the steeple, underground in the crypt, in the echo chapel – little has survived. The government is trying though to restore the monastery. All throughout the Zapata insurrection renovation was halted as the group prohibited the work. I drew a young Indian who stood at the gate of the monastery. Marched on after a brief rest. Deeper and higher into the high forest. We passed the old aqueduct and reached a ridge. Here was a magnificent view. Valleys and mountains. Villages and white paths. Everything glittering in the sun. I couldn't fight off the impression that I was in the Schwarzwald (Black Forest). A snack of bread and oranges, with much ambience and in the company of two soldiers on guard duty in the forest. (Mehring 2016, 97–98)

Figure 3: Reiss sent postcards by the German Mexican photographer Hugo Brehme with popular scenes from Mexico with a view of Tepotzotlán to his wife and friends. This postcard by Brehme depicts the former monastery of Desierto de los Leones.[8]

The *Mexico Diary* shows that Reiss had a clear idea of what he was looking for and what kind of emotional response he expected from his trip south of the U.S. border.

8 This postcard was, however, not sent by Reiss.

After all, he references immediately preconceived images from movie theaters that triggered a certain response after entering Mexican territory. No doubt, he knew about Alexander von Humboldt's travels to Mexico and was familiar with the Mexican imaginary in German cultural history. To a certain degree, the arrival in Mexico and the encounters with Zapatista soldiers, descendants of Aztec people, ancient ruins, colorful markets, and so on, functioned as a self-fulfilling prophecy. The encounter with the rural life of Mexico and people who were not yet conditioned by the routines of American metropolitan lifestyles reveals Reiss's desire to reconnect with his childhood experience in the rural area of the Black Forest. The *Mexico Diary* documents how powerful memories of belonging overcame Reiss during various moments of his trip.

A photograph shows him drawing a Zapatista soldier. On the back of the photograph, Reiss contextualizes the situation by creating a link between the scenery in Mexico with one in Germany. Considering that it is likely that the film was processed after his return to the U.S., the memory of the specific emotion must have been particularly intense.[9] On the back of the photograph, Reiss recognized a striking similarity: "On a mountain close to Mexico City, I drew two soldiers on duty. The area is similar to the Black Forest. Fir trees and moss and then again tropical plants" (Reiss Archives, Reiss 1901, n. pag).[10] In the sketch *Candelario Segura, Mucoloac* offers a lush green background evoking the tall Black Forest fir trees he admired as a young man.

This rare opportunity to look not only at the portrait but also at the very moment in which the portrait is being created, offers us a glimpse into the intercultural encounters which are inscribed in the drawings. With the different media at hand – diary text, drawings, and photographs – we can offer a revealing glimpse at life writing about people who have historically been forced into subordinate positions, exploited and oppressed, and how they contributed to new social developments or how social developments contributed to new forms of (re-)presenting the self and others and other related issues, as the editors of this book explain. The German artist in black jacket, white shirt, and bow tie sitting in front of a revolutionary soldier who is interrupting his guard duty in a forest near Mexico City appears like an echo of the young Winold Reiss drawing rural people in Goeschweiler in the Black Forest.[11] In his Mexican portraits, Reiss brings out the very qualities he cherished in the farmers of the Black Forest. A few years later, in portraying African Americans in Harlem he conflates his visual celebration of

9 This photo is from his Erika Lohmann's album. She was one of his students at the time he was in Mexico. He most likely sent the photo to her and wrote the comments on the back.

10 The drawing he produced is titled *Candelario Segura, Mucoloac, D.F.*, 1920.

11 For a description and analysis of the Black Forest drawings, cf. Mehring (2014, 162–65).

German rustic life with his dedication to the beauty of Mexican folklore, American Indian traditions, and African American aspirations of gaining recognition.

Figure 4: Photograph of Reiss drawing a Mexican soldier with Schaefer sitting next to him.

Figure 5: Winold Reiss, *Candelario Segura*, Mucoloac, D.F., 1920.

It is revealing to compare the act of drawing and writing a diary in Mexico to a bicycle tour he as a fifteen-year-old boy undertook together with his older brother Hans in the area between Freiburg im Breisgau, Kirchzarten, and Triberg im Schwarzwald. The dark firs created a dramatic backdrop which inspired him to write and draw himself into this landscape of his youth. His manuscript "All Heil" from 1901 begins with a sensual description of the looks and flavors of the Black Forest: "The Black Forest, which attracts many foreigners year after year due to its romantic valleys, picturesque houses and high mountains, has helped many a sick person to recover due to its spicy fir air, is also not unknown to the cycler's

Figure 6: Winold Reiss, "Triberger Wasserfall" in his manuscript "All Heil" from 1901. Reiss Archives.

life" (unpaginated first page, Reiss Archives, 1901).[12] This written account and drawings are an early proof that Reiss responded aesthetically to the pleasure of rural life that invited the traveler to ponder the larger implications of life and identity.

In the days following his recognition of a feeling of "Heimat,"[13] Reiss produced powerful portraits such as *Brothers-in-Arms at Zapata's Headquarters in Cuernavaca, Mexico* or *Aztec Indian from Tepotzotlán, Mexico*,[14] which would be reproduced widely in newspapers, magazines, and travel books (see Figures 7 and 11). Today, the drawing of Zapatista soldiers is recognized as "one of the earliest treatments, by an artist of any nationality, of the Mexican Revolution" (Oles 1993, 70). On October 19th, Reiss arrived in Cuernavaca where about 3,000 soldiers resided. Overwhelmed by the sight of ferocious-looking fighters and the vitality of the city, he felt a rush of creativity, as he explained in his diary:

> It is a strangely colorful picture, to see these figures promenade on the plaza of an evening. Music plays. The whole folk is alive and takes joy in these sounds, forgetting the troubles which daily life brings. The Zapatista general was there too. A young, decadent-looking fellow without chin, but with pistol all the larger. Mixed blood of course, like all their leaders, the Indians themselves are peaceful folk as expressed in their being. I drew lots of Zapatistas and was nearly made a general. I also immortalized a Capitan. (Mehring 2016, 98)

An earlier entry, written on October 16th, is a strikingly philosophical reflection on the pettiness of national hatred, German imperialism, and the bloody history of the Mexican Revolution. Reiss concludes: "If we do not build altars to Love now, as the world is falling to pieces, then all is lost, then we go the way many have gone before us, down into chaos, and another, better race than the white race will rise" (Mehring 2016, 124).

Reiss's visual rendering of the Zapatista soldiers combines aesthetics with references to recent historic events of military conflicts in Mexico. He uses bright yellow, red, and blue colors to bring a striking contrast to the melancholic countenance of the armed soldiers. As a result, they appear more like tender brothers than the ferocious bandits he mentions in the diary. Visually, Reiss emphasizes the suffering and humanity in the fighters' posture and expression.

12 It is likely that the Schwarzwald diary was a collaboration between Winold and his brother Hans Reiss with the former providing the drawings and the latter writing the text.

13 The German term *Heimat* is difficult to translate since it connotes a complex conflation of a geographic space and nostalgic memories and fantasies of the place where one feels at home.

14 In his manuscript, Reiss spells the name of the town "Tepozotlan" (no accent) which is close to the spelling "Tepozotlán" used on a Hugo Brehme postcard in 1920. Today's spelling is "Tepotzotlán."

Figure 7: Winold Reiss, *Brothers-in-Arms* at Zapata's Headquarters in Cuernavaca, Mexico. Pastel on paper, 14 x 19 ¾". 1920.

The portrait *Aztec Indian from Tepotzotlán, Mexico* has been criticized as an example of the condescending attitude often assumed by Americans talking about Mexican Indians. For example, James Oles identifies the deliberately unfinished parts of Reiss's drawing below the neck of the Aztec as a white costume, which Americans call 'pajamas.' According to Oles, this kind of clothing represents one of the "innumerable legal impositions forcing the Mexican Indian to abandon his ancient culture" (1993, 75). However, the diary entries about Reiss's four-day-long sojourn in Tepotzotlán reveal a different context. Reiss engaged in conversations with an old captain (Señor Capitan Romero 'with Spanish grandezza') about Mexican history and asked fourteen people to sit for him. Despite the extreme poverty in which Aztec descendants lived, Reiss felt overwhelmed by the warm-hearted spirit which surrounded their personality.

The memories of "Heimat" and repeated references to the Black Forest in the diary offer a clue to understand how Reiss's appreciation for the rural areas in Southern Germany are projected on the artistic depiction of Mexican people. Take for instance the portrait of *Indian Girl from Tepotzotlán.* We see a dark-haired girl with a simple and self-reflective facial expression. In her hands she holds colorful red flowers creating a contrast between the poverty of the dress with the richness of colors and beauty the Mexican landscape offers. The portrait seems to suggest that the girl is offering her flowers as a gift to the viewer (or painter) of the portrait.

There are striking similarities to his portrait *Girl from Oberprechtal (Black Forest)* (1922) which celebrates the quite pride of a young girl in her traditional dress typical for the region. In the diary, he contrasts the simple life and humbleness of the young people to the colorful life they inhabit and the joy that radiates from their eyes.

> Around us the silent steps of the Indians, who served us. In the evening after work they would always come to us. I showed them games, anything that came to mind. And how gratefully they accepted it all, laughed and enjoyed these hours when someone showed them that they have an equal right to life and enjoyment. . . . I found wonderful types. Genuine Aztecs. . . . What sun shone out of these eyes on receiving these simple gifts! Forgotten are cold and everything else, and only the color and certainty of the sweet possession live in these little hearts so lacking in desires. (Mehring 2016, 111)

Figure 8: Winold Reiss, *Girl from Oberprechtal (Black Forest)*, 1922. Color pencil on paper, 19 × 14" (48.2 × 35.6 cm). Private Collection. Image Courtesy Reiss Archives.

The inspiration Reiss receives from being among rural children and drawing their portraits is captured in a photograph from 1908. The painter is the attraction of the city surrounded by young children who curiously look at his skill of catching their likeness on the drawing pad on his lap. While the photograph is clearly staged with the painter at the center of the gathering, it captures the joy Reiss expressed time and time again being surrounded by children. The background offers a glimpse at the rustic dwelling place. Although the image is in black and white, the dresses of the black forest children are known to be colorful and

Figure 9: Winold Reiss, *Indian Girl from Tepotzotlán.*

bright, something Reiss captured in his portraits such as the *Girl from Oberprechtal.* In a letter to his future wife Henriette, he described the scenes of encountering people from the Black Forest in great detail and with admiration: "jobless men, women, and children wandering the streets" and of discovering "interesting types": "lads with faces like robbers," "half-starved women," and "a sweet little wild Italian girl with dark curls and eyes soft as velvet. She was dirty and dressed in rags, but beautiful in her shabbiness. I should have painted her, the color was everything, but she came only once."[15]

If one compares the drawing *Aztec Indian from Tepotzotlán, Mexico* with Reiss's portraits of leading African American artists of the Harlem Renaissance, it becomes quite clear that Reiss was not interested in capturing a white costume. Rather, the areas below the neck are intentionally left unfinished and show the blank paper. Reiss pays particular attention to skin tones and facial expressions. This technique asks the viewer to concentrate on the person's individuality expressed in the facial features. The unfinishedness does not take away from the political dimension of the portraits. One might argue that Reiss's focus on skin color and facial expression in his portraits of Mexicans, Blackfeet Indians, German Americans, Japanese Americans, Chinese Americans, or African Americans

15 Letter from Winold Reiss to Henriette Lüthy, Freiburg, Apr. 22, 1908; Reiss Archives. For more information on the Black Forest experience, cf. Kennedy (2022, particularly 46–47).

Figure 10: Winold Reiss with Black Forest Children, ca. 1908. Unknown photographer. Image Courtesy Reiss Archives.

challenges U.S. American audiences to rethink issues of racial hatred or ethnic xenophobia, and to critically reassess the transnational dimension behind the motto *e pluribus unum*. The technique became a trademark in Reiss's portraits, not only in his portraits of Germans but also in the portraits of African American artists as the example of Zora Neale Hurston shows.

Many of Reiss's philosophical reflections find their equivalent in five paintings which he titles "Mexican Fantasies." These "fantasies" represent imaginative explorations into the history of the Spanish conquest, rituals, and religious traditions. They indicate Reiss's growing interest in issues of cultural confrontation, guilt, shame, and cultural relativism. When Diego Rivera returned to Mexico from his European sojourn, he turned to mural art to revive the study of ancient Mexican designs and translate them into a new spiritual art of visual narrative. In a conversation with Katherine Anne Porter, Rivera calls for a symbolic depiction of Mexican history. Reiss brings a European sense of symbolism in the tradition of Stéphane Mallarmé and Stefan George to the stained-glass windows he encountered in Mexican churches. What he describes as "Mexican Fantasies" resembles this hybrid of German symbolism and Mexican religious symbolic visual narratives.

Figure 11: Winold Reiss, *Aztec Indian from Tepotzotlán*, Mexico, 1920.

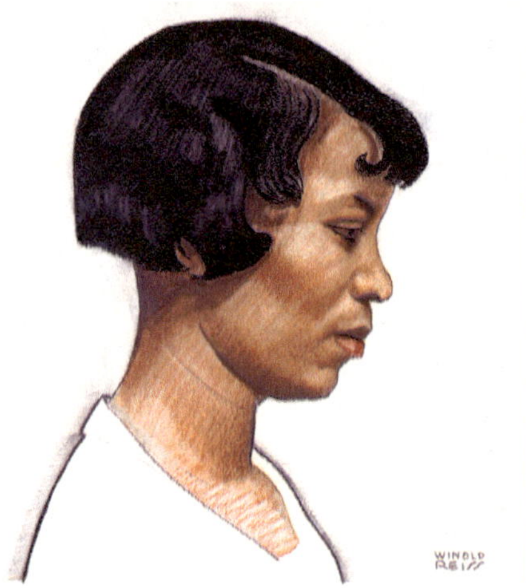

Figure 12: Winold Reiss, *Miss Hurston* (Zora Neale Hurston), 1925.

Reiss gives visual expression to the collision of European imperialism with the Mexican struggle against oppression, using color to infuse his artistic inspirations with a political trajectory. For example, in *Montezuma's Death*, the color white functions as a signifier for instruments of war and death. Purity of sentiment is now associated with the colors Reiss encountered during his trip through Mexico, rather than with white as in religious associations in Western art history.

Mexican Fantasy features two nude figures and a rainbow color line at the very bottom. Reiss is less interested in archaeological accuracy and more in artistic forms, a trend which the photographers Paul Strand, Edward Weston, and Laura Gilpin would emphasize in their work on Mexico a decade later, when the tourist industry had discovered the potential for escapism which the Mexican imagination triggered among big city dwellers. The viewer is invited to cross the Rio Grande with Reiss to trace back the beauty of the abstract color codes in the complexion of the figures, the orange in the trees, the yellow in the rays of the sun, and the green, brown, and blue in the Mexican landscape. In the upper left corner, for example, the rainbow scheme is repeated in a vignette, which opens a window to Mayan rites with stylized figures facing towards the sun.

Mexican Fantasy seems to create a critical dialogue with the color code of *Montezuma's Death*. Here, whiteness as non-color is associated with death and destruction. For Reiss, the Mexican experience undermined the heroic qualities of Wagnerian mythology surrounding figures such as Parsifal or Lohengrin. It revealed a hidden trajectory of brutality and national chauvinism. Reiss shows the white knight as a kind

Figures 13 and 14: Winold Reiss, *Montezuma's Death*, and *Mexican Fantasy III*, both 1920.

of killing machine stabbing naked, defenseless people to death. The lower part of this fantasy evokes a sense of modern-day efficiency and anonymity in the process of mass slaughter. However, above the large skull, which serves as a visual reference to the genocide in the fifteenth century, looms a larger head with the firm profile and proud expression of a survivor. The all-seeing eye with its spiritual power shines on this representative of the Mexican race.[16] The *"Mexican Fantasies"* point directly to Reiss's graphic design elements and to the image *African Phantasy: Awakening*, which he would later contribute to Locke's *The New Negro: An Interpretation*. Like in the *Mexican Fantasy*, we see two naked figures in an idyllic landscape becoming part of the scenery. Masks, rituals, flowers on hills, and clouds resemble repeating metaphors for life, fertility, spirituality, and a connection with ancestry.

During this trip, Reiss experienced an epiphany. Standing amid people from all over Mexico with his board and canvas in front of him, he felt like he had finally

16 Apart from the similarities to the utopian visions of José Vasconcelos, Reiss's Mexican artworks also echo Diego Rivera, who defined his aesthetic theories in his article "From a Mexican Painter's Notebooks." This article appeared in *Arts* in January 1925 as a translation by Katherine Anne Porter. Rivera praised the "Indian aesthetic" for its "pro- found and direct expression of a pure art in relation to the life which produced it." He emphasized that artists needed to convey and feel "the presence of the divine mysterious core hidden within the visible spectacle of the world" (Rivera, qtd. in Walsh and Alvarez 1993, 104).

Figure 15: Winold Reiss, *Mexican Fantasy*, 1920.

Figure 16: Winold Reiss, *African Phantasy: Awakening. The New Negro* (232), 1925.

come "home." The German term *Heimat* [homeland] does not so much represent a concrete location such as the Black Forest or Munich. Neither is there a patriotic sense of "fatherland." Instead, Reiss's term *Heimat* refers to the spiritual realm combined with the gratifying experience of being among people who are very much in tune with their specific folk culture and environment. Before he left Mexico to return to the United States, Reiss conflated a German sense of *Heimat* and *Heimweh* [homesickness] with his Mexican experience: ". . . once again the indescribable Mexican something glimmered in my soul, engraving itself in there, to make me homesick when I am back in the land of materialism (Mehring 2016, 122).

Encountering the rich colors of Mexican culture and the joyfulness of the people, Reiss adds a spiritual component to the notion of belonging, which transcends an actual territory and national borders. Compared to Mexico, the United States appeared to Reiss "dauntingly dead and lifeless" (Mehring 2016, 123). Thus, his trip to Mexico represents not so much an expatriation but a reconnection with something he had lost during his stay in New York City.[17] It is also a multi-

17 The statement is in line with the sentiment of another cultural mediator between Mexican art and American culture, who had left her home in New York City's Greenwich Village and entered

media story-telling that showcases the creative potential of life narratives as suggested by the editors such as the interplay between different forms of media, new connections between self and society, self and environment, and regarding reflections and constructions of the social.

Mediating the Mexican experience in New York City: exhibitions, magazines, murals

In the early 1920s, Reiss was able to make prolific use of the many sketches he had produced during his trip to Mexico. An exhibition at the Anderson Galleries in New York City from November 13–25, 1922, provided a summary of his dedication to the New World, connecting it with his European background in the German and Swedish portraits folio. Reiss's special sensitivity towards the beauty of these people, whose liberating sense of simplicity and joyful cultural conduct had been overlooked, had a great appeal to North American city dwellers. The title of the exhibition offers a remarkable mixture of transatlantic folklore: *The Passion Players of Oberammergau. Drawings by Winold Reiss. Also Drawings from Sweden, Black Forest, Mexico, and Portraits, Woodcuts, and Decorations*. The 154 paintings, fantasies, and woodblock prints presented cultural confrontations and challenged the at the time prevailing concepts of cultural assimilation and "Americanization."

Among the fifty Mexican images on display were portraits, landscapes, fantasies, representations of Mexican street life, textile paintings, and decorations. The exhibition was well timed. In 1922, a traveling Mexican popular arts exhibition under the auspices of the Secretary of Industry, Commerce, and Labor of Mexico, Don Xavier Guerrero, helped generate widespread interest in Mexican culture, traditions, and art. The exhibition was accompanied by a catalogue entitled *Outline of Mexican Popular Arts and Crafts*, edited by Katherine Anne Porter. She provided a detailed overview of the Mexican popular arts and crafts, thereby preparing the

the space south of the Rio Grande at the same time: Katherine Anne Porter. She explained: "I went to Mexico because I was not going into exile, but I was going back to a place I knew and loved" (Porter, qtd. in Brinkmeyer 1993, 29). Like Reiss, she argued in a letter to the editor of *Century* "Why I Write about Mexico" that "New York is the most foreign place I know" (33). Yet, both artists knew that their artistic calling could only come to fruition by working in the urban center where they could function as a bridge between the southern landscape and the cityscapes.

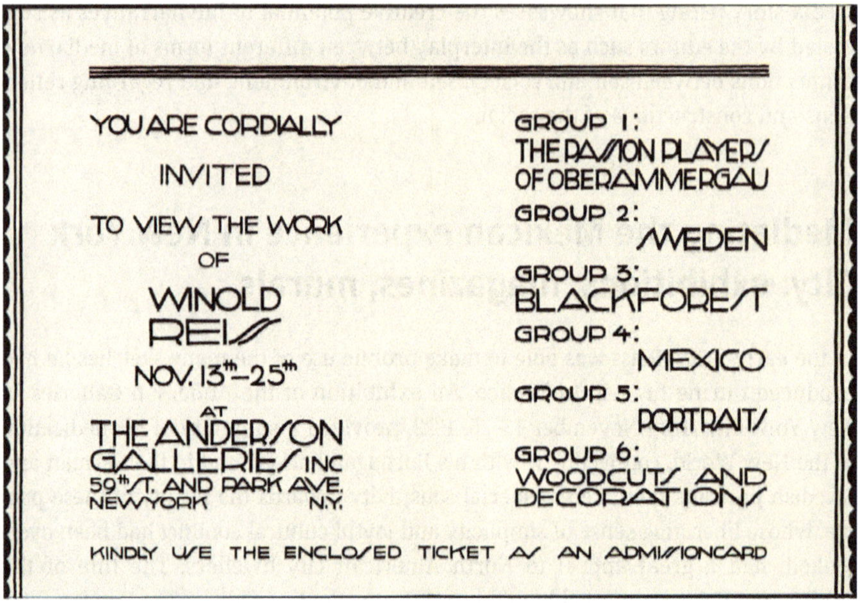

Figure 17: Winold Reiss, Anderson Gallery exhibition folder (inside), 1922.

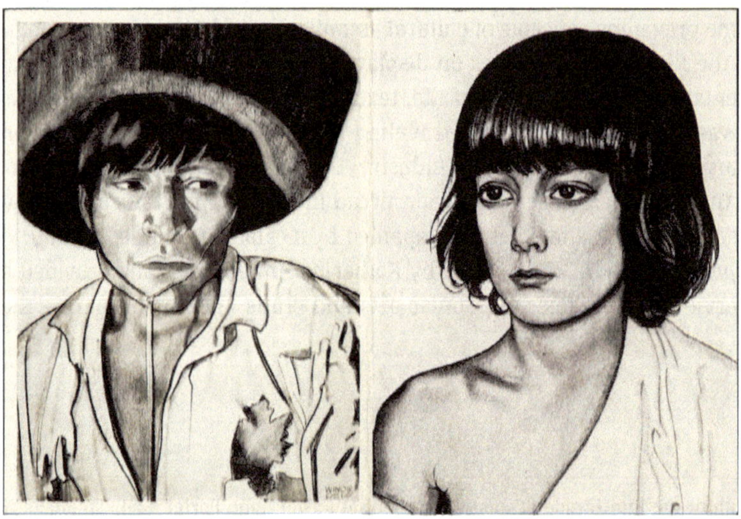

Figure 18: Winold Reiss, Anderson Gallery exhibition folder (outside), 1922.

public for a positive reception of the cultural mediations of Winold Reiss, Diego Rivera, Frida Kahlo, Miguel Covarrubias, and others.[18]

Figure 19: Winold Reiss, *Funeral in Guadalupe*, 1920.

Figure 20: Winold Reiss, *Teoloyucan, Mexico*, 1920.

During the early phase of aesthetic interest among U.S. American travelers in the culture and traditions of Mexico after the revolution, Reiss's Mexican portraits and street scenes appeared in several newspapers and magazines such as *The Century* and *Survey Graphic*. The personal connection between Reiss and the writer Katherine Anne Porter would be crucial for the successful publication, distribution, and mediation of the Mexican drawings. The *Mexico Diary* reveals how and when the two artists met "The next day, Sunday [December 5] I took Best's girlfriend, Miss Porter, and Mrs. Habermann to Tepotzotlán. It was a sunny, wonderful, Mexican day" (Mehring 2016, 121).

18 Sydelle Rubin-Dienstfrey has painstakingly traced the relationship between Reiss and Covarrubias, arguing that they were among "the first visual artists to engage themselves in New Negro representations and physically work in Harlem" (2022, 290). She expanded this approach by including the work of Franz Boas in her recent article on "Transcultural Readings of Winold Reiss, Franz Boas, and Miguel Covarrubias." See Martha J. Nadell's reading of Covarrubias and his association with illustrating "Harlem" in *Enter the New Negroes* (2004).

In the early 1920s, Porter emerged as an important cultural mediator between Mexican folklore and the American literary market. Shortly after she met Reiss on their walking tour to Tepotzotlán, she wrote several stories and sketches about Xochimilco including "In a Mexican Patio" for the *Magazine of Mexico* (April 1921), "Xochimilco" for the *Christian Science Monitor* (May 31, 1921), and "Children of Xochitl" (unpublished, ca. March 1921). She describes this town as an "Island of the Blest" and the people as living in perfect harmony with a beautiful and benign nature. As Thomas F. Walsh has argued, "Xochimilco [expressed] her hope in Mexico as the Promised Land" (1992, 35).

When she was asked to edit a special issue on Mexico for the *Survey Graphic*, she remembered the encounter with Reiss and the sketches he had produced in Mexico. Besides seven Rivera murals, ten examples of Mexican children's art, a portrait of Sor Juana Inés de la Cruz, and Best Maugard's drawings, Winold Reiss's portraits of Indians and a painting of Cuernavaca hold a special place in the magazine with the bold title theme *Mexico: A Promise* (May 1924). It is striking that Porter included not only visual examples of Mexican folklore and Mexican painters such as Diego Rivera and Xavier Guerrero but also portraits and urban scenes from a German American artist. She used Reiss as a non-Mexican presence, from New York's

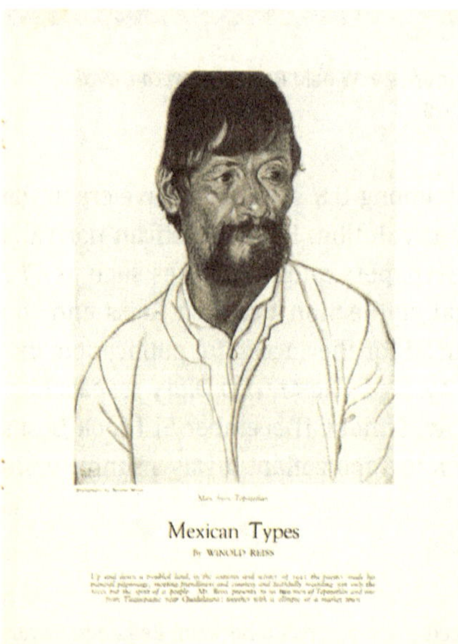

Figures 21 and 22: Winold Reiss, Two "Mexican Types" from *Survey Graphic. Mexico: A Promise* (May 1924).

international art scene, to document the promising qualities inherent in the Mexican art scene emerging after the revolution.

Porter included a section titled "Mexican Types" – similar to the "type" studies, which Reiss would contribute in the following year to Alain Locke's *Survey Graphic* special edition *Harlem: Mecca of the New Negro.* She introduced Reiss's work in the following way:

> Up and down a troubled land, in the autumn and winter of 1921,[19] the painter made his tranquil pilgrimage, meeting friendliness and courtesy and faithfully recording not only the faces but the spirit of the people. Mr. Reiss presents us two men of Tepotzotlán and one from Tlaquepaque near Guadalajara; together with a glimpse of a market town. (Porter 1924, 153)

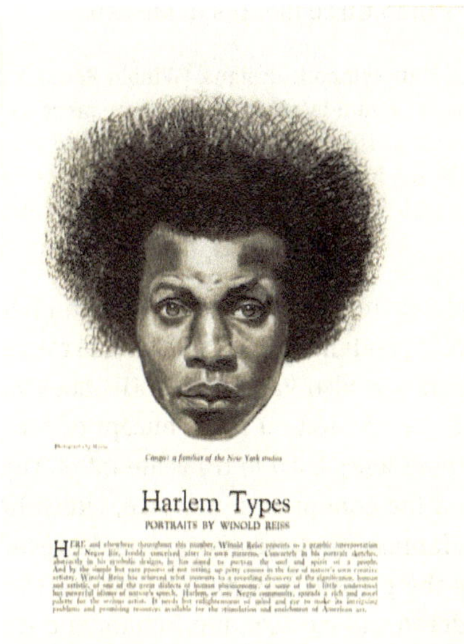

Figures 23 and 24: Winold Reiss, Two "Harlem Types" from *Survey Graphic. Harlem: Mecca of the New Negro* (March 1925).

The expression "pilgrimage" is of particular importance. Porter creates a nexus between Reiss's search for new inspiration and vitality beyond the American urban life of ever-increasing mechanization with her own project of stylizing post-revolutionary Mexico as a promise – not only for its own national future but also for Americans eager to encounter a kind of spirituality from which modern American city dwellers have become estranged. Porter's efforts to draw a highly

19 Porter's date is an error, the trip took place in the autumn of 1920.

persuasive image of Mexican "friendliness and courtesy," the art scene, and the lively vignettes of Mexican folk life captured in her writings stand at the beginning of a soon to be flourishing tourist industry.

With her reference to the kind of faithful dedication, which Reiss's portraits exhibit, Porter implies that the viewer can find a genuine sense of authenticity in the depictions, which former representations of Mexicans were lacking. The combination of counterbalancing allegedly less truthful portraits of non-European people in the Americas and bringing out a sense of spiritual dignity finds an echo in the explanatory notes, which Alain Locke added to the artwork Reiss produced for the *Survey Graphic* special edition on Harlem about a year after Porter's Mexico edition. One might wonder how Reiss could intimately enter the "spirit of a people" in Porter's sense, after spending less than three months in Mexico.

> Concretely in his portrait sketches, abstract in his symbolic designs, [Winold Reiss] has aimed to portray the soul and spirit of a people [sic]. And by the simple but rare process of not setting up petty canons in face of nature's own artistry, Winold Reiss has achieved what amounts to a revealing discovery of the significance, human and artistic, of one of the great dialects of human physiognomy, of some of the little understood but powerful idioms of nature's speech. (Porter 1925, 651)

Porter's description of the colorful marketplaces and the function of Mexican folk ballads in her essay "Corridos" offer a striking pendant to Reiss's visual narrative. Among the contributors to the special edition was also Vasconcelos. His book *La Raza Cósmica* [The Cosmic Race] from 1925 foreshadows a new concept of race, which was very much in line with what Winold Reiss tried to translate into forms and colors on canvas. Vasconcelos predicted the coming of a new age, which he called the "Aesthetic Era." Here, joy, love, fantasy, and creativity would prevail over the rationalism he saw as dominating the present age. Arguing for a fusion of races into a new cosmic race, he considered the *mestizo*, the Indian, and the Black superior to the White race "in a countless number of properly spiritual capacities" (Mehring 2016, 32). In his transnational outline, he insisted that the Iberian part of the continent held the highest promise since the Iberian part possessed

Figure 25: Miguel Covarrubias, Cover, *Survey Graphic: Mexico* (May 1924).

Figure 26: Winold Reiss, *City of Cuernavaca*, 1920.[20]

20 *City of Cuernavaca* appeared in black and white in the May 1924 issue of *Survey Graphic* and in full color in the color gravure section of the *Pittsburgh Sun Telegraph* (June 1, 1930).

"the spiritual factors, the race, and the territory necessary for the great enterprise of initiating the new universal era of Humanity" (38–39).[21]

For the project of racial recognition and turning Harlem into the "Mecca of the New Negro," Reiss's task was to provide portraits and design elements. In a commentary on Reiss's visual work in the *Survey Graphic* issue, Locke praised Reiss as a "folklorist of brush and palette seeking always the folk character back of the individual, the psychology behind the physiognomy" (Locke 1925a, 653). Locke emphasized the spiritual depth of his work, which provided his portraits with an organic structure distinct from what we now call the Art Deco style popular at the time. "In design, he looks not merely for the decorative elements, but for the pattern of the culture from which it sprang" (653). The issue had a print run of more than 30.000 copies and within one year reached an unprecedented number of 42,000 readers. It became a milestone in African American history of cultural self-recognition.

A few months later, the "Harlem" edition was expanded and published under the title *The New Negro: An Interpretation*, including several of Reiss's portraits of leading black intellectuals such as W.E.B. Du Bois, James Weldon Johnson, or Zora Neale Hurston. Locke explained his choice in the following manner:

> The work of Winold Reiss, fellow-countryman of Slevogt and von Reuckterschell [von Ruckteschell] which has supplied the main illustrative material for this volume, has been deliberately conceived and executed as a path-breaking guide and encouragement to this new foray of the younger Negro artists. In idiom, technical treatment and objective social angle, it is a bold iconoclastic break with the current traditions that have grown up about the Negro subject in American art. (1925b, 266)

---

21 If Locke chose Covarrubias to create an international link to the Latin American dimension of racial issues in New York, he and *Survey Graphic* founder Paul Kellogg selected Reiss's artwork to create a cultural bridge to the artistic and philosophical world of Europe. After illustrating the special thematic issue of the *Survey Graphic* on Mexicans, Reiss began his major portrait studies of African Americans and their international heritage in collaboration with Alain Locke. For a detailed exploration of the connection between Reiss and Kellogg, see Patricia Hill's article on "Paul Kellogg, Alain Locke, Winold Reiss, and the Survey Graphic March 1925 issue *Harlem: Mecca of the New Negro*" (2022).

Figure 27: Winold Reiss, *Anton Lang (Christus, Oberammergauer Passionsspiele)*, 1922. Pastel & conté on Bristol paper, 28 1/2 × 19 3/4" (71.1 × 48.9 cm). Private Collection. Image Courtesy Reiss Archives.

Figure 28: Winold Reiss, *Alain Locke*, 1925.

Conclusion

The search for a "new race of great and selfless men" took shape in Mexico, harked back to the experience of living in the Black Forest, and drew Reiss to the issue of the color line (in the Du Boisian sense) back in New York. The *Mexico Diary* and the artwork of Reiss's Mexican experience reconfigure familiar visual narratives of the Harlem Renaissance in a transnational context. Reiss was certainly aware of the strong anti-Mexican sentiments among Americans. Being an immigrant himself, he could well recognize the difference in socio-political status between European and Mexican immigrants. The quality of discrimination and xenophobia suffered by European immigrants during the major waves of immigration in the nineteenth century was different from what historians such as Manuel G. Gonzales described as "the great migration" (2009, 137). The Mexican immigrants were mostly mestizo and Indian who had worked as peasants. In the

1920s, the racial otherness of Mexicans became pronounced in various ways. Gonzales argues that harsh racist views of white Americans functioned as a "justification for the exploitation of Mexicans" and their continual ill- treatment (137). Exhibitions of Reiss's Mexican portraits and his efforts to create a kind of public art, which emphasized a positive perspective on Mexican history, culture, and its people, set a persuasive counterpoint to the dominant Eurocentric perspective of racial superiority.

Reiss criticizes specific developments in modern American urban, industrial societies by positioning Mexican people in an idealized space defined by spiritual integrity, healthy work ethic, patriarch family structures, and colorful sunny landscapes. On canvas, poverty is aestheticized via the beauty of what Reiss describes as pure "Mexican types." The portraits function as a healthy antidote to the despair and uniformity of American modern urban dwellers deprived of joy in the daily routines of working in factories and cramped into housing blocks where sunlight hardly pierces through.

Figure 29: Winold Reiss, *Aztec Woman and Child*, Tepotzotlán, 1920.

Figure 30: Winold Reiss, *The Brown Madonna*. Frontispiece in *The New Negro* (1925).

In New York, Reiss's colorful, joyous, celebratory images left a particularly strong impression. Before the great migration of Mexicans in the 1920s, New York had seen another great migration of African Americans, whose cultural presence had been inscribed in American entertainment in a much more thorough way than that of Mexicans. Aware of the history of slavery, exploitation, lynching, and the struggle for civil rights, Reiss felt drawn to the cause of African American intellectuals in Harlem after the spiritual renewal he experienced during his trip through Mexico. When he ends his diary with a prophetic insight that after thirty-four years, he had finally come to realize in which direction he needed to take his artistry, one might argue that he found in Alain Locke a collaborator who embraced Reiss's dedication regarding the transnational ideal of American togetherness. The image of the Mexican Madonna *Aztec Woman and Child, Tepotzotlán* (Figure 29) finds its counterpart in the *Brown Madonna* on the frontispiece of *The New Negro* anthology (Figure 30). Mexican fantasies and memories of the Black Forest give way to African fantasies. The drawings from Mexico point towards the future when Reiss became dedicated to presenting African American people and culture as central to American history self-representations, they also hark back to his early biography and time in the Black Forest. The image of the Indian Girl finds its counterpart in the *Girl from Oberprechtal* (Figure 8). The spirit inscribed in the faces of Aztecs from Tepotzotlán, the people of the Black Forest, or Oberammergau Passion Players, on display in the Anderson Gallery after Reiss's return from Mexico to New York City, becomes the basis for the iconographic portraits of African American intellectuals and artists.

Epilogue

The bio vignette about Reiss's travels through Mexico as an example of prismatic life writing has been an example to use the gateway of Location as a means to refract the experience into connections to the Black Forest, Munich, the Atlantic crossing, New York City, as well as Social Networks, Media, and Art. The Mexico experience and Reiss's admiration for the culture of Indigenous People has also shaped his idea of using his art to create themed spaces for people in urban American centers. Reiss's architectural and interior design work ranges from his stunning mosaic murals at the Cincinnati Union Terminal to restaurant interiors in New York City – Restaurant Crillon and Longchamps Restaurants among them – where he pioneered a concept of a walk-in *Gesamtkunstwerk*, creating spaces where all design elements were aesthetically coordinated: from thematic murals, ornaments, and decorative elements to tables, chairs, direct and indirect lighting, and menu cards.[22] Art does not only belong in museums but should meet and inspire people in their everyday lives, said Reiss. The American public craves more color, he said, and he would dare to bring color to bear in unprecedented ways. The Tea Room at the Crillon Restaurant in New York City shows how Reiss used his work on Mexican Fantasies with the references to a rainbow color theme, rituals, spirituality, masks, and an idyllic dreamlike landscape to inspire people and encourage them to transcend their urban daily routines. Reiss called it the "prismatic room." This could be a clue for the next bio vignette starting from art and the idea of designing spaces as a "walk-in *Gesamtkunstwerk*" such as the "Congo Room" as an echo of the Wagnerian concept for his opera house in Bayreuth.

Works cited

Bender, Thomas. 2013. "Intellectual Biography and the Matrix of Creativity." In *American Lives*, ed. Alfred Hornung, 3–26. Heidelberg: Universitätsverlag Winter.

Brehme, Hugo. 1923. *Picturesque Mexico: The Country, the People, and the Architecture*. Mexico City: Porrúa-INAH.

Brinkmeyer, Robert H. 1993. *Katherine Anne Porter's Artistic Development: Primitivism, Traditionalism, and Totalitarianism*. Baton Rouge: Louisiana State University Press.

Crawford, Morris De Camp. 1922. The Passion Players of Oberammergau: Drawings by Winold Reiss. Also Drawings of Sweden, Black Forest, Mexico and Portraits, Woodcuts and Decorations. Exh. cat. Anderson Galleries, November 13–25, New York: The Galleries.

Crawford, Morris De Camp. 1921. "Draughtsmanship and Racial Types: Mexican Character Studies by Winold Reiss." *Arts and Decoration* 15, no. 1 (May): 28–30.

Gonzales, Manuel G. 2009. *Mexicanos: A History of Mexicans in the United States*. 2nd ed. Bloomington: Indiana University Press.

Grant, Marion. 1997. *Diverse Reflections: Portraits by Winold Reiss*. Exh. cat., Pittsfield, MA: Berkshire Museum.

Hills, Patricia. 2022. "Paul Kellogg, Alain Locke, Winold Reiss, and the Survey Graphic March 1925 issue *Harlem: Mecca of the New Negro*." In *The Multicultural Modernism of Winold Reiss 1886–1953: (Trans)National Approaches to his Work*, ed. Frank Mehring, 120–39. Berlin: Deutscher Kunstverlag/DeGruyter.

Hornung, Alfred, ed. 2013. "American Lives: Preface." In *American Lives*, ix-xvii, Heidelberg: Universitätsverlag Winter.

Hornung, Alfred. "Anthropology and Life Writing." In *Encyclopedia of Life Writing. Autobiographical and Biographical Forms*, ed. Margaretta Jolly, 2 vols., vol. 1, 38–41. Chicago and London: Routledge.

22 C. Ford Peatross has been a pioneer in mapping and contextualizing Reiss's work in design (2021, 33). See also his article in the book *The Multicultural Modernism on Winold Reiss* where he traced and established Winold Reiss's visual vocabulary in design, drawing, painting, printing, and advertisement (Peatross 2022, 218).

Kennedy, Julie. 2022. "From the Black Forest to New York City: Winold Reiss's Artistic Background in Germany and his Formative Years in Munich." In *The Multicultural Modernism of Winold Reiss 1886–1953: (Trans)National Approaches to his Work*, ed. Frank Mehring, 42–89. Berlin: Deutscher Kunstverlag/DeGruyter.
Kushner, Marilyn Satin, ed. 2021. *The Art of Winold Reiss. An Immigrant Modernist*. New York: New York Historical Society.
Locke, Alain, ed. 1925a. *Harlem: Mecca of the New Negro*. *Survey Graphic* (March).
Locke, Alain, ed.. 1925b. *The New Negro: An Interpretation*. New York: Albert and Charles Boni.
Mehring, Frank. 2014. *The Democratic Gap: Transcultural Confrontations of German Immigrants and the Promise of American Democracy*. Heidelberg: Winter.
Mehring, Frank. 2016. *The Mexico Diary: Winold Reiss between Vogue Mexico and Harlem Renaissance*. An Illustrated Trilingual Edition with Commentary and Musical Interpretation. Trier: WTV, [with Audio CD].
Mehring, Frank, ed. 2022. *The Multicultural Modernism of Winold Reiss 1886–1953: (Trans)National Approaches to his Work*. Berlin: Deutscher Kunstverlag/DeGruyter.
Nadell, Martha J. 2004. *Enter the New Negroes: Images of Race in American Culture*. Cambridge, MA: Harvard University Press.
Oles, James. 1993. *South of the Border: Mexico in the American Imagination, 1914–1947*. Washington: Smithsonian Institution Press.
Peatross, C. Ford. 2022. "Multicultural and Modern Design: The Singular Vision of Winold Reiss, 1906 to 1952." In *The Multicultural Modernism of Winold Reiss 1886–1953: (Trans)National Approaches to his Work*, ed. Frank Mehring, 218–69. Berlin: Deutscher Kunstverlag/DeGruyter.
Peatross, C. Ford. 2021. "Stepping Out in Winold Reiss's New York, 1915 to 1952." In *The Art of Winold Reiss. An Immigrant Modernist*, ed. Marilyn Satin Kushner, 29–48. New York: New York Historical Society.
Porter, Katherine Anne. 1924. "Mexican Types by Winold Reiss." *Survey Graphic* (May): 153.
Porter, Katherine Anne. 1925. "Harlem Types. Portraits by Winold Reiss." *Survey Graphic* (March): 651–54.
Porter, Katherine Anne, Xavier Guerrero, Roberto Turnbull, México Secretaría de Industria, Comercio y Trabajo. 1922. *Outline of Mexican Popular Arts and Crafts*. Los Angeles: Young & McCallister.
Reiss, Winold. 2016. *The Mexico Diary: Winold Reiss between Vogue Mexico and Harlem Renaissance*. An Illustrated Trilingual Edition edited by Frank Mehring with Commentary and Musical Interpretation. Trier: WTV [with Audio CD].
Reiss, Winold. "All Heil." 1901. Manuscript of a bicycle tour of 1901 through the Black Forest. The Reiss Archives.
Reiss, Winold. 1997. *Native Faces: Winold Reiss*. Exh. cat., with an essay by John Heminway. Bozeman, MT: Thomas Nygard Galery.
Reiss, Winold, and Paul Raczka. 1968. *Winold Reiss: Portraits of the Races*. Exh. cat. Russel Museum, Great Falls, with a text by Paul Raczka and introduction by Ned Jacob. Great Falls, MT: C.M. Russell Museum.
Reiss, Winold, Scott J. Tanner, and Richard V. West. 2000. *Winold Reiss: Native American Portraits*. Exh. cat., Seattle: Frye Museum.
Rubin, Sydelle Iris [Rubin-Dienstfrey, Sydelle]. 2002. "Emigrants in Harlem: New Perspectives on Miguel Covarrubias and Winold Reiss." Ph.D. Thesis, Boston University.
Rubin-Dienstfrey, Sydelle. 2022. "Transcultural Readings of Winold Reiss, Franz Boas, and Miguel Covarrubias: Art and Ethnography of the Harlem Renaissance." In *The Multicultural Modernism of*

Winold Reiss 1886-1953: (Trans)National Approaches to his Work, ed. Frank Mehring, 106–19. Berlin: Deutscher Kunstverlag/DeGruyter.
Sampson, P.W. (1931). "A Modern Cellini – Winold Reiss." *Du Pont Magazine* 25.3 (March): 4–6.
Stewart, Jeffrey C., ed. 1989. *To Color America: Portraits by Winold Reiss*. Washington, D.C.: Smithsonian Institution Press; National Portrait Gallery.
Stewart, Jeffrey C. 1990. *Winold Reiss: An Illustrated Checklist of his Portraits*. Washington, D.C.: Smithsonian Institution Press; National Portrait Gallery.
Thoreau, Henry David. 1995. *Walden. An Annotated Edition*. Foreword and Notes by Walter Harding. Boston, New York: Houghton Mifflin Company.
Vasconcelos, José, and Didier T. Jaén. 1997. *The Cosmic Race: A Bilingual Edition*. Balti- more: Johns Hopkins University Press.
Walsh, Thomas F. 1992. *Katherine Anne Porter and Mexico: The Illusion of Eden*. Austin: University of Texas Press.
Walsh, Thomas F., and Ruth M. Alvarez, eds. 1993. *Uncollected Early Prose of Katherine Anne Porter*. Austin: University of Texas Press.

Susana Rocha Teixeira

The Makeover Genre, Life Writing, Men, and Masculinities

Much has been written about life writing in the United States so far. Be it about the blurred lines between fact and fiction in U.S.-American life writing (cf. Adams 1990), the relationship between life writing and self-help literature (Birdsall 2014, 1), the relationship between *Bildungsroman* and life writing (cf. Boon et al. 2023, 20; McLennan 2013, 31), the importance of particular life writing (sub-) genres for American literature and culture and vice versa, or the question of whether the genre of autobiography can be seen as being particularly American, e.g., as being particular compatible and resonating with American ideals and values (cf. Couser 2012, 109–39; Couser 1989; Sayre 1978; McLennan 2013, 29–30, 39). In Couser, for example, it seems that autobiography and memoir can be seen as particular democratic genres that are in concordance with the U.S.-American democratic society and its founding documents (cf. Couser 2012, 26, 31, 120) which purportedly allows all U.S.-Americans to re-make their own lives and selves. Circa one century earlier, W.D. Howells also highlighted the democratic nature of autobiography (Howells 1909, 798). However, it has been well established in academia and elsewhere that neither the founding documents nor these particular life writing genres really meant all individuals living in the United States, since not all were considered (full) Americans or even fully human. As Jim Cullen pointed out, the Declaration of Independence actually meant men, namely "civilized, white males" when it stated that "all men are created equal" and promised as an unalienable Right the "Pursuit of Happiness" (Cullen 2003, 51, 119). Similarly, scholars have argued that the genre of autobiography has been historically primarily preoccupied with male individuals, their life stories, development, as revolving around 'representative' males or representing in one form or another hegemonic masculinity (cf., e.g, Baggerman et al. 2011, 1; Brodzki and Schenck 1988, 1; Civale 2019, 4; Claycomb 2003, 8; Anderson 2001, 3, 34, 43).

Benjamin Franklin's *Autobiography* (1771–1788) has been understood by scholars as being exemplary for the aforementioned. Although critics differ, e.g., in their assessment of why Benjamin Franklin wrote his autobiography, or the reasons for and functions of its reception, there appears to be agreement that his autobiographical writings are of major significance to the genre of autobiography in the United States as well as to the U.S.-American culture and identity itself. It has traditionally been seen as having established the conventions for American autobiography in the aforementioned sense and with it the masculinity "archetype" of the self-made man (cf., e.g., Kimmel 2012, 16). Similarly, Franklin's autobiography has been under-

https://doi.org/10.1515/9783111551968-004

stood as constituting a predecessor to success stories such as Horatio Alger's from rags to riches narratives (cf. Baldellou 2008, 402; Couser, 2012, 118–20; McGee 2005, 31) and pioneering a particular way of presenting, creating, and narrating the American male self and the nation (cf. Couser 1989, 11; Hornung 2019, 1216–219; Lukasik 2018, 12). Following research on the U.S.-American makeover culture (cf., e.g., McGee 2005; Sender 2012), this paper understands the founding father's autobiography as pioneering the makeover genre. It argues that Franklin's autobiography is an early makeover narrative that already uses the key elements and tropes that characterize the makeover (genre) such as the before and after that became so prevalent and scholars primarily identified during the height of the makeover craze in the early twenty-first century.[1]

This essay claims and aims to show, using for example, Franklin's autobiography that the makeover genre is a particular form of life writing that in America gained traction in literature and culture with the emergence of the new nation, and that significantly shaped and participated in U.S.-American conversations and negotiations about male and national self-making and self-invention. Both share notions about narrating (improved) lives, democratization, and (have) historically close ties with hegemonic masculinity, and so-called American values and ideals (cf. e.g., Kavka 2006). In this volatile society that from early on promised rebirth and new beginnings (cf., e.g., Freese 1994, 36, 101, 152; Kimmel 2012, 15, 20; Sender 2012, 29 ff.), the makeover genre fell on fertile ground, which might be one reason for the proliferation of its generic genre conventions that were used and reproduced throughout U.S.-American history. This is not to say that this article advocates American exceptionalism by understanding certain forms of autobiographical writing and makeover narratives as exclusively U.S.-American. These and related notions have been challenged and refuted (cf., e.g., Cahill 2016, 543; McLennan 2013, 10, 30; Graham and Lucero 2020, 126–27). For example, Susan Cahill pointed out the English origins of the "striving and rising" attitude and notions of social mobility that tend to be associated with Franklin and the United States (Cahill 2016, 543). Understanding the makeover genre as a particular form of life writing allows this article to probe and reflect on conventions, functions, and boundaries of both, life writing and makeover narratives.

This article brings different strands of research in life writing studies that has traditionally neglected the makeover genre, research on the U.S.-American makeover culture that traditionally has not particularly been focused on life writing,

1 With his day planner, book of virtues and so on, Franklin contributed to the dissemination of forms of governing the self and the body that are central for makeover narratives, cf. McGee (2005, 6); Reichardt and Schober (2020, 8); Graaff (2020, 81, 88); Danter et al. (2016, 53 ff.).

and masculinities studies into conversation and in doing so highlights the interdependency and co-evolution of particular notions of autobiographical writing, makeover narratives, and masculinity. As personalities such as Frederick Douglass show, self-making and makeover was not limited to males that conformed to or represented hegemonic masculinity. At the same time, his *Narrative of the Life of Frederick Douglass* (1845) or works such as or James Weldon Johnson's *Autobiography of an Ex-Colored Man* (1912/1927) reveal the limits of popular life writing conventions as well as makeover narratives for those who do not conform to hegemonic masculinity.

Making over men

> The makeover is a recognizable format in a country where the dream of self-invention has narrativized new-world colonization since the seventeenth century, where literature from Twain to Dreiser to Fitzgerald has been consumed with the heady rise and sometimes spectacular fall of American Dreamers, and where the 'before-and-after' sales pitch of popular magazines is forever paired in the cultural imaginary with a sand-covered, beached weakling badly in need of a chest-expander. (Kavka 2006)

Much has been written and said in academia about the makeover that appeared to be particularly prevalent in the United States in the early twenty-first century. In numerous makeover TV shows such as *Extreme Makeover* (ABC, 2002–2007), blogs, websites, magazines, novels, movies, TV series, doctor's offices, and elsewhere, U.S.-Americans were improving or making over their bodies, appearances, cars, homes, relationships, and seemingly every other aspect of their lives (cf., e.g., Miller 2008, 3; Weber 2009). Although as numerous examples, e.g., in Arab and Asian countries, or countries in the Americas show, the makeover is certainly not exclusively an U.S.-American phenomenon (cf. Jones 2008, 36–42; Edmonds 2009, 153–70), various scholars nevertheless highlighted makeover's deep roots in American culture and national identity, related notions of self-improvement, new beginnings, and corresponding myths such as the American Dream that have been traced back at least to the seventeenth century (cf. Miller 2008, 2; Heller 2006, 1–6; McGee 2005, 77; Steinhoff 2015, 74–77; Kavka 2006, 212).

According to cultural and media studies scholar Toby Miller, the term makeover seems to have been around since 1860 when it was used in a "satirical magazine" to describe how a "fictional figure . . . through 'miracles of art and culture'" transformed herself into a desirable woman despite her unattractive appearance

(Miller 2008, 1).[2] Yet, following scholars such as historian Kathy Peiss and others, it appears that the makeover narratives, tropes, and aesthetics that seemingly were everywhere in American culture in the twenty-first century, only became more prevalent in the early twentieth century, following a combination of developments such as a growing mass market, or paradigm shifts which, for example, challenged reservations about modifying appearances. This allowed to reframe, e.g., makeup, dieting, or plastic surgery as sensible investments that could increase well-being or facilitate success in the job or the dating market (cf. Peiss 2011,144; Haiken 1997, 7). In the twenty-first century makeover became so ubiquitous in American literature and culture that some critics and scholars called this phenomenon "makeover culture" (cf. Covino 2004, 105; Elliott 2008, 7; Jones 2008,1).

Although a universal, clear-cut definition of what makeover culture is doesn't exist, there seems to be an agreement that in makeover culture individuals recreate and improve themselves and manifold aspects of their lives such as relationships, food habits, psyches, mental capacities, appearances, bodies, pets or houses more or less publicly and ceaselessly (cf., Jones 2008, 1; Steinhoff 2015, 47, 52). Some scholars see makeover culture as being connected to cosmetic surgery culture, others to celebrity culture while other critics understand it in the context of Do-It-Yourself culture, therapeutic or self-help culture (cf. Elliott 2008, 7, 125; Lewis 2008, 1–19, and 2009, 1–4, 7–14; Kavka 2006; Miller 2008, 1–38). Similar holds true for the makeover genre. Although definitions might differ, some of the key elements and tropes that characterize the makeover (genre), and that scholars identified in the early twenty-first century are, e.g., the before-after, the promise of rebirth or new beginning, the notion of hard work, egalitarian optimism, democratization, relations between the inner and outer self, self-improvement and self-making, happiness, upward mobility, exploring and conquering new real or metaphorical frontiers, and the tension between individuality and conformity (cf. Weber 2009; Jones 2008; Steinhoff 2015). These elements and notions are already discernible in Franklin's popular autobiography (cf. McGee 2005, 6–7, 12–13, 31).

Ben Franklin's autobiography has been seen in the tradition of and as being a more secular version of conversion narratives, spiritual growth narratives or spiritual autobiographies aimed at moral or spiritual perfection such as Augustine's *Confessions* or John Bunyan's *Pilgrims Progress*, which was well known and highly significant in early America (Cahill 2016, 544; Couser 1989, 47; Couser 2012, 72, 108–109; Hornung 2019, 1218; McLennan 2013, 44; Seavey 2011, 265, 271; Seavey 1988, 18, 21, 26; Weber 2009, 132). Ben Franklin's autobiography has also been seen in the 'modern' tradition of autobiographical writing that emerged during the Enlighten-

2 On makeover in the mid-twentieth century, cf. Cassidy (2006, 125–39).

ment such as Rousseau's *Confessions* (1782–1789), and as echoing Enlightenment principles and related notions and possibilities regarding the remaking and perfectibility of the individual self (cf., e.g., Arch 2009, 166). However, although Ben Franklin's autobiography does combine numerous elements of these and other then familiar formats of writing (about) the self, it also goes further (cf. Couser 1989, 47). Ben Franklin's autobiography is also a success story of a self-made man who exemplifies upward mobility due to hard work despite his relatively humble beginnings (Baldellou 2008, 402). Ben Franklin's prose played a significant role in creating, disseminating, and establishing this notion or ideal of the self-made man from the eighteenth century onwards. It also embodied the related promises of the American Dream such as upward mobility or agency (many of which are part of makeover culture) and was defined by self-improvement and self-making (cf. Kimmel 2012, 16; McGee 2005, 6; Lukasik 2011, 10). For others, Franklin's narrative laid the groundwork for unfettered materialism and capitalism (cf. Arch 2009, 168). Others saw the model confidence man and other (ethical) issues revolving around the possibilities and limits of remaking oneself in the *Autobiography* (cf., e.g., Lindberg 1982, 73–89). Franklin's autobiography is also considered to be one of the early self-help or advice books. (cf. McGee 2005, 6; Blum 2020, 18; Lukasik 2011, 10; Salazar 2010, 16), which is still read, and occasionally perceived as being one of the best self-help books of all times (cf., e.g., Driscoll 2012, 1; Anderson 2018).[3]

I argue that while the conversion narrative and the other aforementioned categories used so often to classify Franklin's autobiography do not appear to be sufficient to fully describe his prose, seeing Franklin's particular form of autobiographical writing as a makeover narrative might be, because the makeover genre, combines elements of familiar models and modes of narrating the self, which include and go beyond the aforementioned (sub-) genres and notions. For example, the makeover genre also tends to be about (spiritual) transformation, renewal, new beginnings, and the connection between internal and external self. It also provides teleological narratives of progress and success, or about the will and necessity to work hard on the self spiritually and otherwise to realize set goals.

Ben Franklin's autobiography can therefore be seen as an example for how the makeover genre, a combination and continuation of familiar formats of narrating and creating the self, emerged in the Early Republic, during a time in which Americans increasingly left behind the old social order based on kinship and status and entered a dynamic one based on market economy and the malleability of the self

3 McGee understands "self-help literature as a mode of reading, rather than a genre," due to which "nearly any publication – fiction, poetry, autobiography, philosophy . . . could fall within the category" (McGee 2005, 193). I argue that similar holds true for the makeover, which, however, for me is both, a genre and a mode of reading or understanding the source.

(Lukasik 2011, 10,13; Rotundo 1993, 19). Franklin's prose can be seen as being in conversation with earlier forms of male self-fashioning and remaking, as well as with conduct manuals, self-help books, and other cultural products aimed at that purpose that also existed in earlier periods and different regions.[4]

Franklin's autobiography and other works such as Lord Chesterfield's *Letters to His Son* (1774) fell on fertile ground and joined and shaped ongoing conversations in what appeared to be an increasing volatile American society, as they showed aspiring parvenus, who relied on the malleability of the self in order to realize upward mobility and their 'true' or 'desired' selves, or 'calling' (cf., e.g., McGee 2005, 31), how to do it. Here, the manly self-improvement, self-making, and makeover is not only seen as sensible and moral thing to do (cf. Haiken 1997, 7) and it does not merely include hard work, but also work on or management of appearances, because creating the right image of oneself is an important aspect to succeed in the volatile U.S.-American society (cf. Seavey 1988, 62–63, 79–80; Kimmel 2012, 16; Lukasik 2011, 2–5; Cunningham 2012).[5] This climate gave males the opportunity and compelled them to communicate desired ideals and values via their faces, bodies, and appearance in order to re-assert or re-establish masculinity or succeed in the job or dating market (cf. Kimmel 2012; Rotundo 1993). The latest from then on, males felt compelled throughout U.S.-American history to resort to conduct manuals, self-help books, diets or fashion, dying their hair, plastic surgery or other forms of makeover (cf. Vester 2010). This might explain why the makeover's formulaic patterns and generic genre conventions spread throughout U.S.-American history the way they did. However, it was vital that their efforts went unnoticed, in order to prevent losing physical, social or economic capital that tends to be the consequence of visibly or discernably resorting to these activities and displaying interests that tend to be coded as feminine (cf. Weber 2006, 287, 294, 300, 303–304; Weber 2009, 153, 183–184, 201–202, 212).

Making over Ben Franklin

As indicated, this essay takes the notion as jumping-off point that the key elements of the self-made man, makeover narratives, and certain forms of autobiographical writing are similar, interrelated, and in a sense, one. Ben Franklin's

4 Cf. Cahill (2016); Greenblatt (1980) for forms of self-fashioning and self-making in other times and places.

5 Cf. Kimmel (2012, 16) on the importance of creating the right appearance for and in the age of Ben Franklin.

autobiographical writings exemplify this notion. It is also exemplary for how the self-made man was and is created in and disseminated by autobiographies, biographies, novels, and other (sub-) genres that can be seen as forms of makeover narratives. Makeover narratives constitute a form of life writing, which is better suited to describe and categorize works such as Ben Franklin's autobiography than previous alternatives.

What unites self-made manhood, makeover narratives, and certain forms of life writing is, e.g., the possibility of the malleability of the self.[6] Related to this notion of the malleability of the self is the before/after, which is a key element of makeover narratives (cf. Gilman 2001, 36–42; McGee 2012, 686–90; Jones 2008, 7 ff.; Weber 2009, 5–9). Similar holds true for constant monitoring, commenting on, judging, transforming, and improving oneself and every aspect of one's life, including one's mind, food habits, relationships, and body or appearance (cf. Jones 2008, 12). Here, individuals (real or imagined) are constantly striving for perfection to increase happiness, authenticity, or wellbeing, to re-establish their identity (as males) or to realize the American Dream. These aspects became central for makeover TV and other facets of makeover culture (cf. Jones 2008, 12; Weber 2009, 39; Steinhoff 2015, 139–40, 145–46) in the early twenty-first century. Nevertheless, they are already discernible much earlier in Franklin's autobiography.

The oft discussed "errata" in Franklin's autobiography are Franklin's version of recognizing an undesirable or 'deficient' before-state in order to transform this state into a better after-state or what Brenda Weber calls resplendent after-state (cf. Franklin 2012, 25, 43, 45; Weber 2007, 82).[7] As this form of narrating lives fell on fertile ground in the eighteenth century, and even more so in the nineteenth century when the self-reliant, self-made man took center stage, these elements became part of makeover's genre conventions and increasingly re-produced in auto-biographical prose or prose fiction to satisfy the demands of what Benjamin Vaughan had called "a rising people" (Seavey 2011, 268). Like makeover narratives, the biographies, and autobiographies of self-made men, many of which gained particular popularity in the nineteenth century, tend to be success stories that resort to the aforementioned makeover genre conventions. This is why I argue that works such as Ben Franklin's autobiography are makeover narratives that can be considered to be a particular form of life writing.

Although not limited to it, the malleability of the self may also refer to appearances as the notion that creating and communicating the right image of oneself is

6 On the malleability and perfectibility of the self, cf. Danter et al. (2016, 58).

7 On before and after narratives or tropes, cf., e.g., Weber (2007, 82); Lewis (2009, 2, 7). Cf. Kavka (2006, 211–13); cf. Gilman (2001, 36–42), McGee (2005, 686–90).

an important aspect to succeed in the volatile U.S.-American society (cf. Kimmel 2012, 16; Cunningham 2012, 107–08). This notion was also created, disseminated, and established with works such as Ben Franklin's autobiography (cf. Lukasik 2011, 2–5). What is striking is how the description and illustration of Franklin's before-state is reminiscent of how, for example, the before-state of makeover TV candidates is constructed very much later. In both instances the individuals tend to be imagined as being unattractive, unfashionable, not (well-) groomed, questionable hygiene, and of low socio-cultural-economic status (cf. Lukasik 2011, 2; Franklin 2012, 28–29). Furthermore, here, like in makeover TV and other formats of makeover culture the function of highlighting the undesirability of the before-state appears to be "[t]hat you may . . . compare such unlikely Beginning with the Figure I have since made there" (Franklin 2012, 28), i.e., to highlight the miraculous transformation to a magnificent after-state, which is the reward for all the hard work (on the self). For example, in his before-state Ben Franklin is presented as having "a most awkward ridiculous Appearance" for wearing his "working Dress" instead of his "best Clothes," as being "dirty from [his, srt] Journey; . . . . Pockets . . . stuff'd out with Shirts and Stockings." He looks like a lonely, tired, hungry, and impoverished runaway (Franklin 2012, 28) when he first enters Philadelphia and shocks his future wife with his appalling appearance (cf. Lukasik 2011, 2). In contrast, when Franklin returned home after succeeding in his professional life, he "was better dress'd than ever . . ., having a genteel new Suit from head to foot, a Watch, and my Pockets lin'd with near Five Pounds Sterling in Silver" (Franklin 2012, 33), instead of dirty laundry and bread rolls (cf. Lukasik 2011, 2).

Franklin's after-state is not presented as static. Instead, Franklin can adapt, create, and communicate the suitable appearance for each environment or occasion (cf. Lukasik 2011, 5). While his swanky appearance had the function of communicating his success to his brother and potential readers, he attempts to create the right appearance by performing and communicate humbleness and industriousness to his customers (cf. Lukasik 2011, 5; Franklin 2012, 64). For example, Franklin describes that "[i]n order to secure my Credit and Character as Tradesman, I took care not only to be in *Reality* Industrious and frugal, but to avoid all *Appearances* of the Contrary. I dressed plainly" (Franklin 2012, 64; Decker 1997, 138, endnote 45).

The function of the before/after of makeover narratives is therefore to compare and contrast the before and after state to highlight the successful transformation (cf. Marwick 2010, 255; Cardenas 2013, 84, 118), for example, "from the Poverty and Obscurity in which I was born and bred, to a State of Affluence and some Degree of Reputation" (Franklin 2012, 9; cf. Lukasik 2011, 2; Cardenas 2013, 46; Marwick 2012, 255; Weber 2007, 82; Gilman 2001, 37–38). This also points to two other key elements of the makeover that became especially prevalent in the second half of the twentieth century, although already discernible in the nineteenth century, namely the strong relation between inner and outer self, and the promise of an opportunity for new beginnings and leaving a traumatic or otherwise problematic or undesirable past or present behind (cf. Cardenas 2013, 84, 122; Weber 2009 12–13; Sender 2012, 17, 40). Before and after images, narratives, and tropes are some of makeover's most important key elements. Although the representations and functions of the before/after changed throughout time, they seemingly have in common that they illustrate or document the transformation from a deficient before-state or -self into a new or better state or self, and confirm the transformative powers of the makeover, and when applicable, of the experts (cf. Rosen 2004, 19–35). Like a celebrity expert in twenty-first century makeover show, Franklin shares his experiences as author/protagonist, so that others "may find some of them suitable to their own Situations, and therefore fit to be imitated" (Franklin 2012, 9; cf., also Morgan 2007, 134). To conclude: Franklin's *Autobiography* exemplifies why the form of autobiographical writing that he helped popularize can be seen as a makeover narrative, or why this form of makeover narrative can be seen as a form of autobiographical writing. Here hard work and self-discipline are essential for (his) success, but also the creation, management, and improvement of appearances, which are shown to be malleable, and an essential part of his brand and success (cf. Lukasik 2011, 2; Salazar 2010, 63).

Makeover narratives as life writing and advice literature

As indicated before, makeover narratives as forms of autobiographical writing can be seen (or read) as advice literature and as being of particular interest to aspiring parvenus who were trying to remake themselves, to succeed in the job or dating market or realize the American Dream. It is consequently not surprising that a large variety of advice literature was produced, disseminated, and consumed following the American Revolution, particularly during the age of the self-made man in the nineteenth century and after. As Christopher Lukasik pointed out, Lord Chesterfield's *Letters to His Son on the Art of Becoming a Man of the World and a Gentleman* (1774) became highly popular in America and was with at least thirty editions between 1775 and 1800, one of "the most frequently reprinted titles in America"

(Reilly and Hall 2000, 398, qtd. in Lukasik 2011, 56).[8] The *Letters,* which educate about a wide range of topics such as history, literature or politics, can also be seen as an early example of advice literature that taught ambitious men about genteel performances and how to be successful.[9] Similar to Franklin's *Autobiography* one key advice in Chesterfield's *Letters* is that appearances matter. For example, Chesterfield claims that "style is the dress of thoughts; and a well-dressed thought, like a well-dressed man, appears to great advantage. (Chesterfield 1876, 361)." Therefore:

> to be heard with success, you must be heard with pleasure: words are the dress of thoughts; which should no more be presented in rags, tatters, and dirt, than your person should. Do you take great care of your teeth? . . . Are you be-laced, be-powdered, and be-feathered, as other young fellows are, and should be? (Chesterfield 1876, 304)

Not only the correct clothing is essential to create the desired image, which is necessary to succeed, but also the right manners and countenance, which includes the need to conceal one's real sentiments, for example, by controlling or mastering temper (cf. Lukasik 2011, 58). As the following advertisement for Chesterfield clothes illustrates (Figure 1), the name Chesterfield was still used in the early twentieth century to advertise suits, which help "men look as they ought to" as the advertisement says, "and not as they are."

This slogan itself also indicates that Chesterfield's *Letters* was perceived even more than a century later as being about creating a desired image, and not communicating an authentic self. Furthermore, it exemplifies the huge and long-lasting popularity of Chesterfield and his views, which started in the late eighteenth century, in which many works similar to his *Letters* were published and allowed an ever-growing number of men to remake themselves (cf. Lukasik 2011, 17). As scholars pointed out, it is probably no coincidence that as the malleability of the self was advocated by Franklin and others, physiognomy and phrenology started to take center stage in American literature and culture, which presumably allowed, for example, the old elites, to discern a person's (or parvenu's) real character (and heritage) by reading his facial structures (cf. Wegenstein 2012, 8, 23; Gilman 2001, 27, 89–91; Lukasik 2011, 12, 17).

---

8 Chesterfield's *Letters* were published 1775 in the United States (Lukasik 2011, 56).

9 Lewis describes, how, with increasing social mobility, advice literature became increasingly important for the upwardly mobile, who recognized that taste formation, i.e., the performance of distinction via consumption practices regarding home decoration or clothing that signified taste would constitute a form of cultural capital that could facilitate upward mobility. Advice manuals thus "were important at the time in shaping a middle class set of cultural values and identity" (Lewis 2008, 28).

Figure 1: *The Salt Lake Tribune*, Salt Lake City, Utah. November 24, 1906. *The Mission of "Chesterfield" Clothes.* Library of Congress.[10]

I argue in this essay that Ben Franklin popularized the makeover narrative, and with it, a very specific form of life writing that appealed to U.S.-Americans, because the makeover narrative comprises and points to several familiar formats of narrating the self, such as the conversion narrative, while offering advice and egalitarian optimism, for example, regarding upward mobility, to aspiring males. As this form of narrating lives fell on fertile ground in the Early Republic, and even more so in the nineteenth century when the self-reliant, self-made man took center stage, these makeover genre conventions were increasingly re-produced in auto-biographical prose or prose fiction to satisfy the demands of the "rising people." This is, for exam-

10 http://chroniclingamerica.loc.gov/lccn/sn83045396/1906-11-24/ed-1/seq-7/.

ple, discernible in Horatio Alger's from rags to riches narratives, which offer fictionalized versions of and tend to follow the pattern popularized by Ben Franklin, whose autobiography appears to have provided the template for generations of (aspiring) (fictional) self-made men (Leverenz 1989, 75, 83). These works tend to advocate the malleability of the self in various regards, and to resort to familiar conventions established by Franklin and others (cf. Aubry and Travis 2015, 2–5). Although these narratives tend to revolve around the unlikely rise of American Dreamers, about hard work and morals, appearances also tend to play a role in many of these narratives. Some characters may attempt to communicate a certain image by creating a particular appearance that gives them the physical capital they need to, for example, succeed in the job market. In other instances, the new appearances have the function of highlighting the resplendent after-state, i.e., the realized success in the job or dating market.

Horatio Alger (1832–1899) was a commercially very successful author in his time and up until the 1920s (cf. Pitofsky 1998, 279), which can be seen as an indicator for the interest in makeover narratives during that period. *Ben, the Luggage Boy* (1870) from the *Ragged Dick Series*, for example, appears to follow Ben Franklin's model and genre conventions. Ben, the main character, also runs away like his namesake from a relatively respectable home at a young age due to conflicts with his family. Both, makeover narratives and autobiographies claim to offer a particular unmediated access to authenticity or truth. This is also the case in *Ben, the Luggage Boy*, where we learn in the preface that "it is in all essential points a true history" (Alger 1870, pos. 3477), and that the story is based on Ben's "own statement" (Alger 1870, pos. 3477), although the author/narrator appears to sneak in his "own observation" as well, which is, as he appears to suggest nevertheless the truth (Alger 1870, pos. 3477). What this story has in common with other Alger narratives such as *Bound to Rise* (1873) is the centrality of appearance as the commenting on, modifying, and improving of appearances appears to be a staple in Horatio Alger's from rags to riches narratives. Here the right appearance tends not to be presented in a negative way. It is not a sign of vanity or lack of masculinity, but a necessity to succeed in the job or dating market. Consequently, for young males, good looks, nice clothes, a great haircut, and cleanliness are presented as necessary investment or as indicating the resplendent after-stage, as marking a new beginning and success. Ben's good looks and face facilitate his professional and personal success. However, as Ben also learns early on, it is important to resort – like his namesake – to the adequate appearance. At the beginning, Ben's appearance is too neat, which is why he fails to find a job. In accordance with physiognomy's teachings, his appearance also indicates his pedigree (Alger 1870, pos. 3497, 3627). However, the main source of his success is his hard work, increasing frugality (Alger 1870, pos. 5628), and luck of meeting benefactors who helped him. What distinguishes this story from many others and in particular Ben Franklin's makeover narrative, is that the story ends with Ben's return to his family and friends and a complete makeover that was "like the change that transformed Cinderella into a princess" (Alger 1870, pos. 5789).

The success story of the makeover genre – a model for everyone?

Despite its popular success, some criticized the Franklinian way of narrating and re-making the self. For example, Edgar Allan Poe's short story "The Business Man" (1845; cf., e.g., Lemay 1982; Person 2001, Zimmerman 2005, 26),[11] which revolves around Peter Proffit, and is a thinly veiled parody of Franklin's autobiography, and the way Franklin is presented and created in it (cf. Tschachler 2013, 39; Savoye 2018, 102; Hayes 2009, 84). "The Business Man" mimics Franklin's makeover narrative in various regards. For example, the usage of lists or day planners, which are useless in Poe's short story because of Dadaist-like entries such as "July 11. To do. do. do" (Poe 2012, 296). It also echoes Franklin's presumably rational approach to individuals and things. However, in the case of Peter Proffit this kind of thinking boils down to ruthless behavior and is presented as being the result of a kind of mental disability, caused not by a loving mother but a loving nurse who swung him in such a way that he bumped his head on a bedpost (Poe 2012, 294; Schnackertz 1999, 61). In contrast to Franklin's or Alger's makeover narratives, there is no great reveal or resplendent after-state in "The Business Man," where the character presents the new or improved version of himself. Instead of documenting self-improvement, self-making or progress of this runaway, "The Business Man" appears to offer a reverse makeover narrative that documents the protagonist's unmaking as 'businessman' as he moves down from profession to profession until he ends up in the "Cat-Growing" business (Poe 2012, 300; cf. Person 2001, 159), where he chops off cattails and sells them to the government to

11 As John E. Reilly pointed out, various versions of E.A. Poe's "The Business Man" exist. I use the edition that was published in 1945 (cf. Reilly 1999). One of the transformations affects the name of the narrator-protagonist that changed from Pendulum to Proffit, i.e., from con man to making profit. The preceding version was called "Peter Pendulum, the Business Man" (cf. Savoye 2018, 102; Armiento 2018, 126–27; Kennedy 2006, xix). Related issues are discussed in Poe's "Epigram for Wall Street" (1845) (cf. Tschachler 2013, 4, 9, 20 43, 44, 73) and Poe's "Philosophy of Furniture" (1840) (cf. Cunningham 2012, 116).

receive bounty as legislature had passed a law to mitigate the infestation with cats (Poe 2012, 300; cf. Duquette 2007, 11–12). There is no character development, but only what feels to be an endless list of different documentations and occupations. This further exposes Franklinian-style makeover narratives as empty shells of genre conventions, which also holds true for the self-made man (cf. Tschachler 2013, 145). The sardonic "[i]n biography the truth is everything, and in autobiography it is especially so" (Poe 2012, 295) further highlights this notion. There is no truth to makeover narratives, there is no truth to autobiographies, there is no authentic self, especially not, if you're writing about yourself, and the self-made man as proposed by Franklin and others does not exist either (cf., e.g., Tschachler 2013, 72).

The fact that the business man worked as "a walking advertiser" (Poe 2012, 299) further emphasizes that in this world the characters are only defined by their profession, appearance, and other superficialities. Seen this way, "The Business Man" appears to be heralding works such as Bret Easton Ellis's *American Psycho* (1991) where notions of growth and self-improvement that tend be associated with makeover narratives are also absent. Instead, the main character Patrick Bateman, who may or may not be a serial killer, appears to see the world through brands, occupation, and status. Similarly, Peter Proffit once worked as organ grinder, which is an instrument. However, considering the way Peter Proffit was presented in "The Business Man," the term organ grinder evokes an ambivalence and suspicion that it probably would not be beneath Peter Proffit to operate a machine that grinds (animal or perhaps even human) organs for profit. This passage can therefore be seen as a commentary on the nineteenth-century business world where the profit of one person is based on the exploitation of another. This can mean being lied to, robbed, or cheated by shapeshifting businessmen or conmen (although the lines between the two often are blurred), as for example discernible in Herman Melville's *The Confidence Man: His Masquerade* (1857), in Peter Proffit or his 1842 predecessor Peter Pendulum (cf., e.g., Hayes 2009, 84; Cunningham 2012, 108; Tschachler 2013, 16; Hurh 2015, 129; Lemay 1982, 34).

This can also be seen as commentary on the lot of the individual in the nineteenth century workplace, where man must adapt to the company and the market's needs. Here, instead of being an individual, man becomes a function, as it is, for example, also discernible in Melville's "Bartleby, the Scrivener" (1853). As Duquette pointed out, this critique appears to be reminiscent of Emerson's "American Scholar" (1837) that also criticizes "[t]he state of society . . . in which the members have suffered amputation from the trunk, and strut about so many walking monsters, – a good finger, a neck, a stomach, an elbow, but never a man. Man is thus metamorphosed into a thing, into many things" (Emerson 1837, 57; Duquette 2007, 10–11, Lemay 1982, 32).

Writers such as E.A. Poe and Hermann Melville criticized and made fun of the related focus on appearances and superficialities, for example, in "The Man That Was Used Up" (1839), "Spectacles" (1844), and *The Confidence Man: His Masquerade* (1857), respectively. "The Man That Was Used Up" revolves around a narrator/protagonist who appears to be interested in, almost obsessed with Brevet Brigadier General A.B.C. Smith, "a fine-looking fellow" (Poe 2004, 190), who appears to exude an air of importance and mystery (Poe 2004, 190–91). The irony here is that the narrator highlights the "*remarkable* . . . individuality of the personage" (Poe 2004, 191) who turns out to be made of nothing but clothes, prostheses, and rumors.

At first glance, General Smith appears to be what Kimmel would probably call a heroic artisan, a real, authentic man, a man of war who has not yet been effeminated by the malls, luxury, consumption, and the need to remake himself in order to succeed in his professional and private life. He is reminiscent of and appears to be a hyperbolic version of characters such as Colonel Manly, whose function in Royall Tyler's play *The Contrast* (1787) was, for example, to help reflect on which masculinity archetype will be needed and best for the new nation (cf., e.g., Kimmel 2012, 11–14). While Mr. Dimple is presented as effeminate, immoral dandy in *The Contrast* who is interested in appearances and luxury, Colonel Manly is presented, as his telling name indicates, as more manly, because he represents an unmodified and thus uncorrupted and incorruptible, more authentic, or real form of masculinity, which is framed as the better alternative for his love Maria and the nation. The narrator who appears to be obsessed with General Smith in "The Man That Was Used Up," excessively lists the General's perfections. He was tall and with a "commanding" presence (Poe 2004, 191). He has "an *air distingué* . . . which spoke of high breeding and hinted at high birth" (Poe 2004, 191). He has full, flowing, glossy, jetty black hair, "the most entirely even, and the most brilliantly white of all conceivable teeth" (191). His body has "wonderful proportion[s]" (Poe 191). Unsurprisingly, General Smith was "one of the *most* remarkable men of the age" and a "favorite . . . with the ladies" (192), an indicator of his virility and masculinity. The "gallant soldier" is interested in philosophical conversations and welcomes progress and change. He appears to be happy to "live in a wonderful age" with "steam-boats" (197). As the narrator proceeds to know more about the general, in an attempt to learn who he really is, the narrator sees that the general is nothing more than a heap of clothing and prostheses, "a large and exceedingly odd looking bundle of something" (197) with "one of the smallest, and altogether the funniest little voices, between a squeak and a whistle" (197). The wonderful hair is a wig and the teeth are dentures (198). General Smith is helpless and depends on his Black servant, Pompei, who literally makes General Smith, as he makes the General out of the different prostheses. This is a not-so-subtle reminder that the right to make and remake oneself, to realize oneself, to be self-reliant, has been reserved for white males throughout American history. Females, Blacks, and other others have been excluded from this dream as they had the function of helping white males (re)make themselves (McGee 2005, 7, 13). George Washington, for example, had enslaved persons helping him with his appearance, e.g., powdering his hair (cf. Coe 2020). His teeth were also not his. Washington wore dentures that probably were made of teeth by enslaved persons (AP 2020).

"The Man That Was Used Up" appears to challenge makeover narratives by turning them on the head regarding generic genre conventions. Here, instead of transforming a deficient before-state into a resplendent after-state, the opposite is the case. It also makes fun of other simplistic views as propagated, for example, by physiognomy or phrenology as the narrator appears to be convinced to be able to find physical traits that point to the General's fine character and heritage, only to find that the whole general is fake. However, although the narrator seems to be naïve throughout the short story, "The Man That Was Used Up" repeatedly voices suspicions. For example, the name Smith is a very common name which challenges the General's alleged exceptionality. The reference to a sculptor (Poe 2004, 192) or Mr. Smith's stiffness are further indicators of his artificiality. "The Man That Was Used Up" also rejects notions of presumably authentic, unmodified, or real forms of masculinity as embodied by Colonel Manly or General Smith. The short story appears to suggest that if even such an authentic, real male is made, then it could be that the unmodified, uncorrupted male does not exist. Seen this way, "The Man That Was Used Up" as well as "The Business Man" are not only representative of the dark romanticist's (such as Poe or Hawthorne) challenges regarding uncritical views regarding the perfectibility of the self, it also echoes Poe's preoccupation with masculinity, an aspect that he had in common with Emerson.

Other works such as F.S. Fitzgerald's *The Great Gatsby* (1925) or W.D. Howells's *The Rise of Silas Lapham* (1884/1885) also famously pointed to the limits of male self-making (cf., e.g., Decker 1997, 91–92; Marchand 2010, 311). The novel *The Rise of Silas Lapham* revolves around the protagonist Silas Lapham who appears to have realized the American Dream as he rose from rags to riches. He too appears to be reminiscent of characters such as *The Contrast's* Colonel Manly, who represents the heroic artisan, but also of characters such as Mr. Van Rough, *The Contrast's* self-made man.[12] However, in *The Rise of Silas Lapham* it is not the female character

12 Kimmel, *Manhood* (11–14) offers this reading of the characters in *The Contrast*, which allows to explore the different masculinity archetypes. *The Contrast* is also discussed in Lukasik (2011, 64–71).

who has to choose the right man for her and America's future. It is Tom Corey, one of the male main characters, who has to choose between Silas Lapham's daughter Irene, who stands for appearances and beauty and Penelope, who represents knowledge and intelligence. While Tom's father is an example for what Kimmel described as genteel aristocrat, his son appears to be a combination of a businessman as represented by his grandfather and the genteel aristocrat embodied by his father, Bromfield Corey. The title of the novel and the first pages of *The Rise of Silas Lapham* promise a conventional makeover or from rags to riches narrative a la Franklin or Horatio Alger. It appears to follow so much so generic genre conventions that the journalist Bartley Hubbard who interviews Silas at the beginning of the novel, seems to be able to predict the elements of Silas's life story such as Silas's poor childhood (cf., e.g., Murphy 1989, 25). In doing so, the journalist appears to be echoing the sentiments of readers who might feel the same way while getting Silas's advice, and hearing about Silas's wonderful mother, his roots on a rural farm, his hard work, and luck – because it was his father who found the mineral that provided the basis for the very successful paint (Howells 2018, 3–6). However, although *The Rise of Silas Lapham* appears to be reminiscent of a classic Franklinian autobiography or Horatio Alger-like makeover narrative, we are presented with a reverse makeover narrative as Silas Lapham falls from riches to rags, because he loses everything which is also accompanied by a deteriorating appearance.[13]

Although Silas Lapham initially appeared to have become a very successful businessman and to have realized the American Dream, he still faces numerous sources of insecurity. Since his status as self-made man is not based on heritage or other stable sources, but on wealth and success, he inhabits an inherently insecure position since the wealth or success of the self-made man "is as easily unmade as it is made" (Kimmel 2012, 13; also cf. Kimmel 2012 7, 11–14; Armengol 2013, 32). Thus, self-made men have to constantly prove their masculinity, make, remake, improve themselves, and work on themselves in order to succeed, fearful, as Gary Wills writes, that "his product will get out of date, or rot in the storehouse" (qtd. in Kimmel 2012, 7) especially against the background of tough competition. Furthermore, with increasing wealth and luxury, the self-made man was increasingly perceived as effeminate (cf. Kimmel 2012, 78). Furthermore, the novel is quick to reveal that this self-made man is not self-made after all, since Silas Lapham's success is also the result of his father finding the mineral, his wife helping him, and his former business partner who gave him much needed capital.

---

13 Some critics see *The Rise of Silas Lapham* still as a success story, because his economic and social downfall appears to be accompanied by his moral growth. Cf., e.g., Dimock (2010, 84); Lavin (2011,362–82); Li (1996, 113–20).

At the same time, *The Rise of Silas Lapham* shows with Bromfield Corey, which appears to be Howell's version of the genteel patriarch, and the opposite of Silas Lapham, that what appears to be an old aristocratic family, turns out to be a fairly recent one, as it was Bromfield's father who as an old India merchant, made the family rich (cf. Cox 1991, 111).

Colonel Silas Lapham served as union solder in the Civil War and still takes care of the widow and daughter of one of his men. This and his "personal appearance," including what Bartley described as "square, bold chin" and "pair of massive shoulders" underlines Silas Lapham's presentation as heroic artisan who likes to work. At the same time, he is presented as self-made man, as a parvenu without style or etiquette as he does not get up to welcome the journalist and gives the left hand instead of the right to greet him (Howells 2018, 3). This and certain facial markers such as "his forehead," which is "good, but broad rather than high" (4) might give him away as being of low birth to a reader of the novel and connoisseur of physiognomy in the nineteenth century. Silas Lapham appears to be keenly aware of his lower-class markers, which is why he is almost obsessively preoccupied with his appearance, which ultimately heralds his fall. He is constantly reassuring himself, and others, that he and his family are as worthy as members of the old aristocracy. He is constantly monitoring and reflecting on his appearance and performance when he encounters others, trying to make sure that he appears to be on equal footing if not more. Silas Lapham's obsession with appearances is particularly discernable in the "gloves-scene" when Silas Lapham, his wife, and daughter Irene finally meet the Coreys, and enter the lion's den of high society. For various pages (141–43), Silas Lapham obsessively reflects on what he should wear, whether he should buy gloves, and so on. He is so obsessed with his appearance that the text places him in proximity to his wife and daughters, ergo in the feminine sphere. The "gloves-scene" is the turning point in the novel that visibly marks the beginning of the end of Silas Lapham and highlights the reason for his fall. While he is constantly worried about the impression he makes (although suggesting otherwise), his pride prevents him from adapting and modifying himself, and thus from acquiring the social, cultural, and symbolic capital that is as much needed as the economic to really make it. The fact that the aristocrat's son Tom Corey marries Lapham's daughter Penelope, who at the end of the novel has lost her economic capital but still has her cultural capital, further appears to emphasize this point.

The limits of remaking the self

As the examples have shown, the Franklinian kind of self-making and life writing has its limits in many regards. In addition to the here discussed examples, critics have repeatedly highlighted that this kind of self-making, and life writing tends to exclude persons who do not conform to hegemonic masculinity ideals. We know that although autobiography is purportedly a democratic genre (cf. Couser 2012, 128), it was not available in the same way to women, African Americans, and others who did not conform to hegemonic masculinity (cf. McLennan 2013, 4, 19; Huddart 2008, 2–3). As numerous works such as Booker T. Washington's *Up From Slavery* (1901) or Frederick Douglass's *Narrative of the Life of Frederick Douglass* (1845) show, African American upward mobility stories exist. Nevertheless, they might differ significantly in numerous regards. One often discussed aspect is that for many of these (often slave) narratives the end of their journey, e.g., freedom or the ability to educate themselves tends to be the starting point of makeover narratives or forms of life writing as popularized by Franklin. We also know from works such as Monica L. Miller's *Slaves to Fashion* (2009) that appearances also played a significant role in the African American community. It is also widely established that Frederick Douglass was known for his impeccable appearance and was also one of the most photographed males of his time. Nevertheless, the possibilities of self-making were very limited to the majority of African Americans. McGee argued that in the dominant ideology, women and people of color were excluded from the idea of self-making, because the mythology of the self-made man relied on their unacknowledged labor and servitude and the ideal of individual success and self-invention was reserved to white males (McGee 2005, 7). Therefore, and because as stated previously, it appears that (male) self-making, makeover narratives, and certain forms of life writing are interconnected, it is not surprising that these are the most visible ones in American literature and culture, although exceptions do exist. Following current research such as Elizabeth McHenry's *To Make Negro Literature* (2021) that focuses on and excavates forgotten works and actors, there is hope that in the future we will be able to find more works that revolve around female or African American self-making, makeover narratives that have been neglected or forgotten in the past, and therefore able to rewrite U.S.-American history of self-making and re-writing the self.

Works cited

Adams, Timothy Dow. 1990. *Telling Lies in Modern American Autobiography*. Chapel Hill: University of North Carolina Press.

Alger, Horatio. 1870. *Ben, the Luggage Boy. The Collected Complete Works of Horatio Alger Jr.* Amazon.de. Kindle. https://www.amazon.de/Collected-Complete-Collection-Including-Campaign-ebook/dp/B06XF3HZVR.

AP News. "Store yanks souvenir Washington dentures over slavery link." 19 February, 2020. https://apnews.com/general-news-b6a3c672b25352a306539828827b9bd9.

Anderson, Linda R. 2001. *Autobiography (The New Critical Idiom)*. London: Routledge.

Anderson, Tom. 2018. "Benjamin Franklin: The Unlikely Father of Self-Help." *Blinklist Magazine*, January 17. https://www.blinkist.com/magazine/posts/benjamin-franklin-father-self-help.

Armengol, Josep M. 2013. "Embodying the Depression: Male Bodies in 1930s American Culture and Literature." In *Embodying Masculinities. Towards a History of the Male Body in U.S. Culture and Literature*, ed. Josep M. Armengol, 31–48. New York: Peter Lang.

Armiento, Amy Branam. 2018. "Poe in Philadelphia." In *Poe and Place*, ed. Philip Edward Phillips, 123–44. Cham: Springer.

Aubry, Timothy, and Trysh Tavis. 2015. "What is 'Therapeutic Culture,' and Why Do We Need to 'Rethink' it?" In *Rethinking Therapeutic Culture*, ed. Timothy Aubry and Trysh Travis, 1–23. Chicago and London: University of Chicago Press.

Arch, Stephen Carl. 2009. "Benjamin Franklin's *Autobiography*, Then and Now." In *The Cambridge Companion to Benjamin Franklin*, ed. Carla Mulford, 159–71. Cambridge: Cambridge University Press.

Baggerman, Arianne, Rudolf Dekker, and Michael Mascuch. 2011. "Introduction." In *Controlling Time and Shaping the Self. Developments in Auto-biographical Writing since the Sixteenth Century*, ed. Arianne Baggerman, Rudolf Dekker, and Michael Mascuch, 1–10. Leiden: Brill.

Baldellou, Marta Miquel. 2008. "From Boys to Men in Edward Bulwer-Lytton's *Pelham* and Horatio Alger's *Struggling Upward*: The Dandy and the Self-Made Man Coming of Age." *E.H. Filologia* 30: 401–12.

Birdsall, Kate Elizabeth. 2014. *Becoming a Better Me: The Co-Evolution of Contemporary American Memoir and Bestselling Self-Help Literature, 1952–2013*. PhD-thesis. Michigan State University.

Blum, Beth. 2020. *The Self-Help Compulsion. Searching for Advice in Modern Literature*. New York: Columbia University Press.

Boon, Sonja, Laurie McNeill, Julie Rak, and Candida Rifkind, eds. 2023. *The Routledge Introduction to Auto/biography in Canada*. New York: Routledge.

Brodzki, Bella, and Celeste Schenck. 1988. "Introduction." In *Life/Lines Theorizing Women's Autobiography*, ed. Bella Brodzki and Celeste Schenck, 1–15. Ithaca: Cornell University Press.

Cahill, Edward. 2016. "The English Origins of American Upward Mobility; or the Invention of Benjamin Franklin." *ELH*. 83, no. 2: 543–71.

Cardenas, Elaine. 2013. *Quick Change: The Extreme Makeover of American Life*. PhD-thesis, George Mason University.

Cassidy, Marsha F. 2006. "The Cinderella Makeover. *Glamour Girl*, Television Misery Shows, and 1950s Femininity." In *The Great American Makeover. Television, History, Nation*, ed. Dana Heller, 125–40. New York: Palgrave Macmillan.

Civale, Susan. 2019. *Romantic Womens Life Writing. Reputation and Afterlife*. Manchester: Manchester University Press.

Claycomb, Ryan M. 2003. *Playing at Lives: Life Writing and Contemporary Feminist Drama*. PhD thesis. University of Maryland.
Coe, Alexis. 2020. “Five myths about George Washington. No, he didn’t wear a wig, and he wasn’t a great military commander.” *The Washington Post*. https://www.washingtonpost.com/outlook/five-myths/five-myths-about-george-washington/2020/02/13/eed41ea8-42e2-11ea-b503-2b077c436617_story.html.
Couser, G. Thomas. 1989. *Altered Egos. Authority in American Autobiography*. New York: Oxford University Press.
Couser, G. Thomas. 2012. *Memoir. An Introduction*. New York: Oxford University Press.
Covino, Deborah Caslav. 2004. *Amending the Abject Body: Aesthetic Makeovers in Medicine and Culture*. State University of New York Press.
Cox, James M. 1991. “The Rise of Silas Lapham: The Business of Morals and Manners.” In *New Essays on The Rise of Silas Lapham*, ed. Donald E. Pease, 107–28. Cambridge: Cambridge University Press.
Cullen, Jim. 2003. *The American Dream: A Short History of an Idea that Shaped a Nation*. Oxford: Oxford University Press.
Cunningham, Patricia A. 2012. “Fashion, Furnishings, and Style.” In *Edgar Allen Poe in Context*, ed. Kevin J. Hayes, 107–17. Cambridge: Cambridge University Press.
Danter, Stefan, Ulfried Reichardt, and Regina Schober. 2016. “Theorising the Quantified Self and Posthumanist Agency Self-Knowledge and Posthumanist Agency in Contemporary US-American Literature.” *Digital Culture & Society* 2: 53–70.
Decker, Jeffrey Louis. 1997. *Made in America. Self-Styled Success from Horatio Alger to Oprah Winfrey*. Minneapolis: University of Minnesota Press.
Dimock, Wai-Chee. 1991. “The Economy of Pain: Capitalism, Humanitarianism, and the Realistic Novel.” In *New Essays on The Rise of Silas Lapham*, ed. Donald E. Pease, 67–90. Cambridge: Cambridge University Press.
Driscoll, Molly. 2012. “10 best self-help books of all time. 1. ‘The Autobiography of Benjamin Franklin,’ by Benjamin Franklin.” *The Christian Science Monitor*. April 26. https://www.csmonitor.com/Books/2012/0426/10-best-self-help-books-of-all-time/The-Autobiography-of-Benjamin-Franklin-by-Benjamin-Franklin.
Duquette, Elizabeth. 2007. “Accounting for Value in ‘The Business Man.” *Studies in American Fiction* 35, no.1: 3–30.
Edmonds, Alexander. 2009. “‘Engineering the Erotic’: Aesthetic Medicine and Modernization in Brazil.” In *Cosmetic Surgery: A Feminist Primer*, ed. Cressida J. Heyes and Meredith Jones, 153–70. Farnham: Ashgate.
Elliott, Anthony. 2008. *Making the Cut: How Cosmetic Surgery Is Transforming Our Lives*. London: Reaktion Books.
Franklin, Benjamin. *Autobiography (A Norton Critical Edition)*. 2012. Ed. Joyce E. Chaplin. New York and London: Norton.
Freese, Peter. 1994. *“America” – Dream or Nightmare? Reflections on a Composite Image*. Essen: Die Blaue Eule.
Gilman, Sander L. 2001. *Making the Body Beautiful. A Cultural History of Aesthetic Surgery*. Princeton: Princeton University Press.
Graaff, Kristina. 2020. “Racialized Self-Improvement: Advice in Black and White Self-Help of the Interwar Years.” In *Laboring Bodies and the Quantified Self*, ed. Ulfried Reichardt and Regina Schober, 81–102. London: Routledge.

Graham, Maryemma, and Mercedes Lucero. 2020. “Life Writing.” In *The Routledge Handbook to the Culture and Media of the Americas*, ed. Wilfried Raussert, Giselle Liza Anatol, Sebastian Thies, Sarah Corona Berkin, and José Carlos Lozano, 121–35. London: Routledge.
Greenblatt, Stephen. 1980. *Renaissance Self-Fashioning: From More to Shakespeare*. Chicago: Chicago University Press.
Haiken, Elizabeth. 1997. *Venus Envy. A History of Cosmetic Surgery*. Baltimore: Johns Hopkins University Press.
Hayes, Kevin J. 2009. *Edgar Allan Poe*. Reaktion Books.
Heller, Dana A. Ed. 2006. *The Great American Makeover. Television, History, Nation*. Basingstoke: Palgrave Macmillan.
Heller, Dana A. Ed. 2006. “Before: ‘Thinks Just Keep Getting Better.’” In *The Great American Makeover. Television, History*, Nation, ed. Dana Heller, 1–7. New York: Palgrave Macmillan.
Hornung, Alfred. 2019. “North America.” *Handbook of Autobiography / Autofiction*. Martina Wagner-Egelhaaf, ed., 1205–260. Berlin: DeGruyter.
Howells, William Dean. 1909. “Editor’s Easy Chair.” *Harper’s Monthly Magazine*. October 1: 119, 713; ProQuest pg. 795.
Howells, William Dean. 2018. *The Rise of Silas Lapham (A Norton Critical Edition)*, ed. Paul R. Petrie. New York and London: Norton.
Huddart, David. 2008. *Postcolonial Theory and Autobiography* (*Routledge Research in Postcolonial Literature)*. London: Routledge.
Hurh, Paul. 2015. *American Terror. The Feeling of Thinking in Edwards, Poe, and Melville*, Stanford: Stanford University Press.
Jones, Meredith. 2008. *Skintight: An Anatomy of Cosmetic Surgery*. Oxford: Berg.
Kavka, Misha. 2006. “Changing Properties: The Makeover Show Crosses the Atlantic.” In *The Great American Makeover. Television, History, Nation*, ed. Dana Heller, 211–30. New York: Palgrave Macmillan.
Kennedy, Gerald J. 2006. “Introduction by Gerald J. Kennedy.” In *The Portable Edgar Allan* Poe, ed. Gerald J. Kennedy, ix-xxx. London. Penguin Books.
Kimmel, Michael S. 2012. *Manhood in America. A Cultural History*. Third edition. New York: Oxford University Press.
Lavin, Matthew J. 2011. “Clean Hands and an Iron Face: Frontier Masculinity and Boston Manliness in *The Rise of Silas Lapham*.” *Western American Literature* 45, no. 4: 362–82.
Lemay, JA Leo. 1982. “Poe’s *The Business Man*: Its Contexts and Satire of Franklin’s ‘Autobiography.’” *Poe Studies* (1971–1985) 15, no. 2: 29–37.
Leverenz, David. 1989. *Manhood and the American Renaissance*. Cornell University Press.
Lewis, Tania. 2008. *Smart Living. Lifestyle Media and Popular Expertise*. New York: Peter Lang.
Lewis, Tania. 2009. “Introduction.” In *TV Transformations: Revealing the Makeover Show*, ed. Tania Lewis, 1–6. London: Routledge.
Lewis, Tania. 2009. “Changing Rooms, Biggest Losers and Backyard Blitzes: A History of Makeover Television in the United Kingdom, United States and Australia.“ In *TV Transformations: Revealing the Makeover Show*, ed. Tania Lewis, 7–18. London: Routledge.
Li, Hsin-Ying. 1996. “For Love or Money: Courtship and Class Conflict in Howells’ *The Rise of Silas Lapham*.” *Studies in American Fiction* 24, no.1: 101–21.
Lindberg, Gary H. 1982. *The Confidence Man in American Literature*. New York: Oxford University Press.
Lukasik, Christopher. 2011. *Discerning Characters: The Culture of Appearance in Early America*. Philadelphia: University of Pennsylvania Press.

Lukasik, Christopher. 2018. "Illustrating A Life: Remediating Franklin's Autobiography in Nineteenth-Century America." In *Intermediality, Life Writing, and American Studies: Interdisciplinary Perspectives*, ed. Nassim Winnie Balestrini and Ina Bergmann, 9–35. Berlin: De Gruyter Mouton.
Marchand, Mary. 2010. "Faking It: Social Bluffing and Class Difference in Howells's *The Rise of Silas Lapham." The New England Quarterly* 83, no. 2: 283–12.
Marwick, Alice. 2010. "There's a Beautiful Girl Under All of This: Performing Hegemonic Femininity in Reality Television." *Critical Studies in Media Communication* 27, no. 3: 251–66.
McGee, Micki. 2005. *Self-Help, Inc. Makeover Culture in American Life*. Oxford: Oxford University Press.
McGee, Micki. 2012. "From Makeover Media to Remaking Culture: Four Directions for the Critical Study of Self-Help Culture." *Sociology Compass* 6, no. 9 (September): 685–93.
McLennan, Rachael. 2013. *American Autobiography (BAAS Paperbacks)*. Edinburgh: Edinburgh University Press.
Miller, Toby. 2008. *Makeover Nation. The US of Reinvention*. Columbus: Ohio State University Press.
Morgan, Jeff. 2007. "The Founding Father: Benjamin Franklin and his Autobiography." *Romanticism and Parenting.* In *Image Instruction and Ideology*, ed. Carolyn A. Weber, 128–37. Newcastle: Cambridge Scholars Publishing.
Murphy, Brenda. 1989. "Howells and the Popular Story Paradigm: Reading Silas Lapham's Proairetic Code." *American Literary Realism, 1870–1910* 21, no. 2: 21–33.
Peiss, Kathy. 2011. *Hope in a Jar. The Making of America's Beauty Culture*. Philadelphia: University of Pennsylvania Press.
Person, Leland S. 2001. "Poe and Nineteenth-Century Gender Constructions." In *A Historical Guide to Edgar Allan Poe*, ed. J. Gerald Kennedy, 129–65.
Pitofsky, Alex. 1998. "Dreiser's The Financier and the Horatio Alger Myth." *Twentieth Century Literature* 44, no. 2: 276–90.
Poe, Edgar Allen. 2004. "The Man That Was Used Up." In *The Selected Writings of Edgar Allen Poe (Norton Critical Editions)*, ed. Gary R. Thompson, 190–99. New York and London: Norton.
Poe, Edgar Allen. 2012. "The Business Man." In *Benjamin Franklin's Autobiography (A Norton Critical Edition)*, ed. Joyce E. Chaplin, 294–301. New York and London: Norton.
Reichardt, Ulfried, and Regina Schober. 2020. "Introduction: Laboring Bodies and the Quantified Self." In *Laboring Bodies and the Quantified* Self, ed. Ulfried Reichardt and Regina Schober, 7–14. London: Routledge.
Reilly, John E. 1999. "The 'Missing' Version of Edgar Allan Poe's 'The Business Man.'" *American Periodicals* 9: 1–14.
Rosen, Christine. 2004. "The Democratization of Beauty." *The New Atlantis. A Journal of Technology & Society* 2, no. 5 (Spring): 19–35.
Rotundo, E. Anthony. 1993. *American Manhood. Transformations in Masculinity from the Revolution to the Modern Era*. New York: Basic Books.
Salazar, James B. 2010. *Bodies of Reform: The Rhetoric of Character in Gilded Age America*. New York: New York University Press.
Savoye, Jeffrey A. 2018. "Poe and Baltimore." In *Poe and Place*, ed. Philip Edward Phillips, 97–122. Cham: Springer.
Sayre, Robert F. 1978. "Autobiography and the Making of America." *The Iowa Review* 9, no. 2: 1–19.
Schnackertz, Hermann Josef. 1999. "Of Bumps and Brains: E.A. Poe and Phrenology." In *Lost Worlds & Mad Elephants. Literature, Science and Technology 1700–1990*, ed. Elmar Schenkel and Stefan Welz. Glienicke and Berlin: Galda and Wilch.
Seavey, Ormond. 1988. *Becoming Benjamin Franklin. The Autobiography and the Life*. University Park: Pennsylvania State University Press.

Seavey, Ormond. 2011. "'The Manners and Situation of a Rising People': Reading Franklins Autobiography." In *A Companion to Ben Franklin*, ed. David Waldstreicher, 255–74. Malden: Wiley-Blackwell.
Sender, Katherine. 2012. *The Makeover. Reality Television and Reflexive Audiences*. New York: New York University Press.
Stanhope, Philip Dormer, Earl of Chesterfield. 1876. *Letters Written by the Earl of Chesterfield to His Son*. Philadelphia: J.B. Lippincott & Co.
Steinhoff, Heike. 2015. *Transforming Bodies. Makeover Monstrosities in American Culture*. Basingstoke: Palgrave Macmillan.
Tschachler, Heinz. 2013. *The Monetary Imagination of Edgar Allan Poe: Banking, Currency and Politics in the Writings*. McFarland.
Vester, Katharina. 2010. "Regime Change: Gender, Class, and the Invention of Dieting in Post-Bellum America." *Journal of Social History* 44, no. 1 (Fall): 39–70.
Weber, Brenda R. 2006. "What Makes the Man? Television Makeovers, Made-Over Masculinity, and Male Body Image." *International Journal of Men's Health* 5, no. 3 (Fall): 287–306.
Weber, Brenda R. 2007. "Makeover as Takeover: Scenes of Affective Domination on Makeover TV." *Configurations* 15, no. 1: 77–99.
Weber, Brenda R. 2009. *Makeover TV: Selfhood, Citizenship, and Celebrity*. Durham: Duke University Press.
Wegenstein, Bernadette. *The Cosmetic Gaze: Body Modification and the Construction of Beauty*. MIT Press, 2012.
Zimmerman, Brett. *Edgar Allan Poe: Rhetoric and Style*. McGill-Queen's Press-MQUP, 2005.

Part II: **Creativity**

Wilfried Raussert

Playing it Cool, or when Experience Meets Creativity: 'An Aesthetics of the Cool' in Life Writings by Olaudah Equiano and Zora Neale Hurston

Beyond testifying to authorship and humanity of the black subject, Olaudah Equiano's *The Interesting Narrative of the Life of Olaudah Equiano, or Gustavus Vassa, the African* (1794) and Zora Neale Hurston's "How It Feels to Be Colored Me" (2020 [1928]) express unique personal life experiences. Clearly, context and experience differ in Equiano's and Hurston' s case. Equiano wrote in the context of transatlantic abolitionist struggle, Hurston spun her narrative during the heydays of the New Negro movement that propelled black pride in Harlem in the 1920s. What they share are hemispheric experiences of the black condition. Equiano, as traveler and 'travelee,' became acquainted with the African, Caribbean, and North American cultural context. Hurston, as anthropologist and folklorist, got exposed to the American South, the North of the U.S., as well as Jamaica and Haiti where she investigated the remnants of African and Voodoo cultures in the Caribbean. She published her results in *Tell My Horse: Voodoo and Life in Haiti and Jamaica* (1938).

Looking at autobiographical writings by Olaudah Equiano and Zora Neale Hurston, this essay departs from the hypothesis that life writing represents an ideal textual space where experience and creativity meet and shape one another. Life writing, as I argue in this essay, provides an opportunity authenticating an expression not only of the self but in particular of the experience of the self in relation and context. With a nod to Mita Banerjee, I view the above texts as centering on "the notion of lived experience" (Banerjee 2018, 359). While experience is frequently dismissed as too abstract a category in poststructuralist thinking championing the discourses of cultural construction and representation, I consider individual experience, particularly of black selves, as fundamental in coming to terms with the creation of self in life writing texts.

Let us assume that the generally expected collective narrative about black cultures in the Americas would embark from questions like "What could be harder than creating a sense of self when you were not even master of your own body, 'when even the body was not legally one's own'" (Alexander 2019, 190)? Enslaved, expropriated, dispossessed, and exploited, black people were responsible for their survival, endurance, and liberation; for cultural and social creativity against all odds, for social resistance and establishing an aesthetic vanguard

https://doi.org/10.1515/9783111551968-005

in situations of control, surveillance, and oppression (Butler and Athanasiou 2013). It takes a strong will to endure and an immense gift of creation to overcome the adverse circumstances of much of black colonial history and succeed.

The above collective narrative clearly expresses a quintessential colonial experience. However, it frequently ignores the significance of lived black experience. Equiano's and Hurston's texts, as I argue, literally give life to individual experience. Not only do they manifest black authorship, but also narrate from a perspective of black agency and individualized black subjectivity. They may be puzzling to the reader precisely because they cherish the author's own creative potential of performativity relating the self to agency in the context of community, migration, and encounter with the other. As I suggest in this essay, as their writings move between and beyond fact and fiction, Equiano and Hurston provide visions of black subjectivity beyond mere representation.

What is more, in my reading, their agendas as autobiographical authors are not concerned with a reproduction of a colonial collective narrative of black oppression and exclusion. While such narrative hovers in the background of their texts, more so in Equiano's than Hurston's text, these writers discover in life writing the creative use of performativity as a means of narrating black experience differently. Looking at black history through the eyes of a playful, performative life writing self, they spin narratives of individual experience adding different perspectives and "alternative knowledge" to the discourse of black experience as they further expand the spectrum of black social relations (Banerjee 2018, 324). They do so by effectively bringing together Western and African traditions of thinking and writing, thus creating transcultural expressions of lived experience.

As the autobiographical writings by Equiano and Hurston show, we need to be aware that, despite shared colonial experiences, black cultures and the narratives about them differ. Whether the black population occupies a minority or majority status within national boundaries marks, for instance, significant differences between African Americans and Africans. Comparing African American and Black British communities, Marie Hélène Laforest emphasizes that "within the black diaspora, despite claims to global brotherhood and sisterhood – not that solidarity between blacks conceptually excludes diversity – there is a constellation of blackness, different ways of being black. This was as true in the past as it is today" (Laforest 1996, 115). In particular, she points out historical differences between the United States and England which have rendered different shades of consciousness among their black populations.

Olaudah Equiano and Zora Neale Hurston migrate and write in different cultural, economic, and historical contexts. We may speculate at this point that the creation of self and identity in Equiano's text is different from its counterpart in Hurston's essay, partly due to individual and historical differences, partly due to the fact that different conditions call forth the performance of empowered black subjectivity. In which way Equiano's and Hurston's performances of mobile and empowered black selves actually differ and in which way they also overlap can only be answered after an analysis of performativity and transformation in the texts themselves. Reading the selected life writing texts in their transcultural dimension, I maintain that the African concept of 'thinking and acting cool' shapes the expression of poly-local lived experience by both writers. While critical readings of Equiano's and Hurston's writings tend to emphasize elements of fluidity and hybridity in the context of Western autobiography, transatlantic relations, and African American cultures, they downplay African philosophical ways of thinking and acting creatively that these texts are produced upon.

With that in mind, it appears a logical consequence to see Equiano's and Hurston's writings as transculturally inspired texts that can best be interpreted in a broader frame of life writing than within the boundaries of more conventional Western autobiographical modes of writing. Highly transcultural in their selection of narrative, performative, and ethnographic practices, both texts, as I argue, demonstrate a playful blackening of identity politics, of narratives of self, and of knowledge production based upon African philosophical thinking and the expression of lived experience.

An aesthetic of the cool

In his essay "An Aesthetic of the Cool," Robert Farris Thompson explores "the understanding of a basic West African/Afro-American metaphor of moral aesthetic accomplishment, the concept cool" (1973, 41). In most African and black diasporic cultures, the metaphorical connotations of 'cool' signify control, composure, and social security. As Thompson explains, 'cool' is frequently associated with sacred rituals of a purifying nature involving the use of water and chalk. Hence, 'cool' refers to "a return to freshness, to immaculate concentration of mind, to the artistic shaping of matter and societal happening" (41). 'Cool' is connected to black philosophical thinking which shares the belief "the cooler a person becomes, the more ancestral he (she) becomes" (41). Important for the act of the creation of self in relation to the social, the successful practice of the philosophy of 'cool' leads to a mastery of self. Transcending time and space, "he can concentrate or she can concentrate upon truly important matters of social balance and aesthetic substance, creative matters . . . ." (41). Frequently, coolness takes on the role of a mask that is put on in times of trouble as well as exaltation in black dance, performance, and arts. As I argue in this essay, 'playing it cool' turns into an autobiographical mode of writing that empowers the black self in a performative way and creates new narratives different from the colonial, Black Atlantic, and enslavement discourses in which we see black writings predominantly embedded. The self repeatedly poses as cool, calm, and collected in friendly and hostile settings in Equiano's and Hurston's texts. We may think of Equiano's acting as a parson in the context of a funeral and Hurston's entertaining performances for passing tourists in Eatonville, Florida. Writing the self from within an aesthetic of the cool, Equiano's and Hurston's life writing texts create a supreme agency and express individual mastery of performing empowered selves.

Their life writings celebrate creativity, inventiveness, and performative control that play with black and white dichotomies and cultural differences only so that they ultimately transcend them. 'Acting cool' becomes a means of shaping the social and cultural environment from the base of a well-balanced and simultaneously creative self that employs individual experience as a foundation for new social visions.

Recent trends in transcultural life writing studies

Whereas "[f]or a long time, the classical forms of autobiography as full-blown and complete accounts of human lives were considered the only version of the genre worthy of discussion" (Hornung 2013, x), the latter half of the twentieth century showed several new developments in the studies of autobiography and biography which manifested themselves in new perspectives on a traditional genre and the now common usage of the term 'life writing' to describe different forms of self-expression. Life writing can be studied from different perspectives, ranging from literary studies to history, philosophy, and psychology. Indeed, much of recent critical work on race, gender, class, sexual orientation, as well as disability, particularly in regard to postcolonial politics and aesthetics, focuses on life-writing discourses to discuss issues of identity, memory, history, agency, and representation (Moore-Gilbert 2009; Walder 2011).

This trend clearly continues in the twenty-first century as the expansion of texts belonging to the rather broad category of life writing as well as criticism surrounding the phenomenon have continuously flourished and gained recognition (Smith and Watson [2001] 2010; Hornung 2009a and b, 2013). Life writing has "developed into a privileged form of self-expression and into a major field of scholarship" (Hornung 2009b, 536) and as such, it has turned into a field which is in persistent need of revaluation in terms of definition and methodology since concepts of self and narrative are never fixed.

What most contemporary critics in these fields agree upon is that life writing is not a single or unified form of writing. Rather, it should be viewed as an expression which encompasses several genres such as autobiography, memoir, slave narrative, letters, testimonio, court reports, and so on (Egan and Helms 2004; Smith and Watson [2001] 2010; Banerjee 2018) and as Marlene Kadar posits, "a critical practice" on the part of the ideologically charged reader (1992, 3). Moreover, contemporary life-writing studies, and narrative studies more generally, put up for debate distinctions of fact and fiction (Maftei 2013) as life narrators often use complementary modes of fact and fiction consciously to intertwine history, personal, and collective memory, borrowed recollections, as well as myth – a practice suggested by terms such as "faction," "autofiction" and "autobiographical novel." In this way, narrators are able to piece together fragments of past and present, as well as imagination and (historical) knowledge/reality to produce narratives which are characterized by multi-perspectivity.

In regard to a working definition, this essay aligns itself with Sidonie Smith and Julia Watson who in their seminal work *Reading Autobiography: A Guide for Interpreting Life Narratives* ([2001] 2010) employ the term life writing as a "general term for writing that takes a life, one's own or another's, as its subject. Such writing can be biographical, novelistic, historical, or explicitly self-referential and therefore autobiographical" (Smith and Watson 2010, 4). Most important for the relation between text and the social, life writing texts are distinguishable in regard to "their relationship to and claims about a referential world" (10). Life narrators, as opposed to novelists, "anchor their narratives in the world of their own temporal, geographical, and cultural milieu" (10). Alfred Hornung argues that "[m]ore than a representation of self, such texts represent their subjects in relation to their cultures, tracing their interactions over periods of time" (2009b, 536). And, as I would add, telling and narrating life as an experience of self in relation to others opens ways to discover life writing as an expression of empowering the self as creative and transformative voice that challenges grand narratives (Lyotard).

This becomes particularly relevant for this essay as it focuses on the life writing discourses of transnational and migratory experience. For Alfred Hornung, the practice of life writing "has become a form of intercultural negotiation with the goal of realizing a transcultural form of existence" which "challenge[s] beliefs in national allegiance and geographical boundaries" (2009b, 537). Similarly, Smith and Watson have argued that in life-writing discourse, "narratives . . . have long been sites through which formerly marginal and displaced ethnic and racialized subjects explore terms of their cultural identities and their diasporic and transnational allegiances" (2010, 156). Each of these narratives, according to Lisa Brown, may be viewed as having a particular "performative function" (2011, 276). Performativity based on lived experience and balanced by an 'aesthetic of the cool' is what the essay at hand explores in Olaudah Equiano's and Zora Neale Hurston's life writings that rethink self and social relations beyond black resistance.

Olaudah Equiano's *The Interesting Narrative of the Life of Olaudah Equiano, or Gustavus Vassa, the African, Written by Himself*

The Interesting Narrative of the Life of Olaudah Equiano, or Gustavus Vassa, the African, Written by Himself (1794) is generally regarded as the first major slave autobiography in American literature. On the level of content and form, it served as a prototype for numerous slave narratives used by abolitionist groups to fight slavery prior to the Civil War. Moreover, narrative patterns as developed by Equiano have been relocated in texts such as *Narrative of the Life of Frederick Douglass, An American Slave* (1845) and Harriet Ann Jacob's *Incidents in the Life of a Slave Girl* (1861). But the impact of Equiano's autobiography also permeates fiction and non-fiction in the twentieth century such as Richard Wright's *Black Boy* (1945) and *The Autobiography of Malcolm X* (1965). In spite of such clear embedment of Equiano's text within the realm of African American literature, one needs to be aware, though, that his text could also be considered a vital contribution to Afro-British and Afro-Caribbean literature since Olaudah Equiano's narrative reflects upon several migratory experiences between Africa, America, the Caribbean, and Europe. As slave, servant, free trader, and tourist, he narrates a search for and performance of identity beyond clear-cut national, racial, and cultural alliances. In this work, a sense of the individually experienced moment at a certain site and a specific time in history becomes crucial for narrating the self. From the experience of self in context emerge larger social visions of culture and economy and reflections on coexistence.

As Geraldine Murphy reminds us, "many critics have noted that slave narratives are transcultural or hybrid productions" (2001, 369–70). She stresses the hybrid textual quality of Equiano's narrative when she states that "[h]is narrative incorporates elements of spiritual autobiography, the newly emerging secular success story, and the political discourse of the humanitarian/abolitionist movement, not to mention travel writing" (369). Indeed, much of Equiano's narrative is based upon extensive reading in the antislavery and travel literature of his time. Also, Equiano infuses his autobiography with a Franklinesque spirit of free-enterprise, of self-reliance, and a call for work ethic and frugal practice. However, what critics seem to have missed to point out is Equiano's strong indebtedness to African culture and African philosophical thinking. An aesthetic of the cool, balancing and harmonizing seemingly opposite forces, shapes Equiano's performativity of self right from the start of the book. Already, the frontispiece illustrates Equiano's eagerness to stage his self-conception as a composite African-European-American individual. What we encounter is a hybrid and perplexing image. While the portrayal reveals striking African features, Equiano's attire emphasizes the social status gained in British society. He actually poses as a self-made man of letters; thus highlighting his authorship. The dress code becomes an expression of social and cultural achievement and sets the tone for the narrative of empowerment to follow. The European attire especially helps underscore Equiano's pride about his social standing. At the same time, the frontispiece appears to be ambiguous, an effect that is further intensified by the double-naming on the very title page. While the African name Olaudah Equiano and the European name Gustavus Vassa represent only two of various names adopted by or forced upon him, they clearly highlight an awareness of self even beyond DuBoisian double-consciousness on part of the author. While Equiano appears to be aware of his ambiguous position within the European world, he claims, at the same time, a certain degree of ease with feeling at home in multiple affiliations. In all this, a sense of shifting identities – dependent on the experience of time and place – is further intensified when the name Gustavus Vassa is immediately followed by the reference to the continent of his origin on the frontispiece.

Equiano's at times playful self-conception as hybrid, migratory subject suggests that his look at the world is not a unified one. Similar to the absence of a clear, definitive cultural grounding in his life and narrative (autobiography as oral genre has a long history in African cultures as well), his perspectives on cultures shift. Equiano, overall, seems to share the contemporary paradigm viewing Europe as colonizer and Africa and the West Indies as the colonized.[1] Nevertheless, underneath such comparative hierarchical opposition, there exists a subtext in Equiano's narrative which leaves the reader uncertain about the status of cultures in the author's thinking. Equiano praises European culture and sees educational and economic patterns as models for imitation. And yet, he also refers to Westerners as "white savages" (2001, 40) in particular when he describes their

1 Angelo Constanzo explains that "[t]he influence of 18th century primitivism on the slave narrative was strong. Several of the autobiographers, such as Gronniosaw, Cugoano, Equiano, and Smith, recalled their memories of idyllic days that ceased when European influence spread over Africa. These writers knew of the powerful sentimental attraction that such ideas as primitivism and the noble savage had on the eighteenth century Western world, and thus the narrators described their memories of Africa according to the prevailing primitivist notions popularized by the numerous travel books written by autobiographers and historians" (2002, 350).

cruel treatment of black slaves but also when he criticizes their lack of cleanliness and his dislike for some of their everyday habits. Similarly, his perception of non-Western cultures denies a monolithic view. At times, he detects moral and democratic values that are far superior to the ones he encounters in European cultures; at other times he emphasizes the profits non-Western cultures would gain from colonization.

Equiano's attitude towards Africa reveals a similar pattern of ambiguity. He clearly identifies with his place of birth but also distances himself from cultural practices at the same time, as surprising pronoun shifts in the text illustrate. When he describes the everyday life of his African tribal environment, he leads the reader into an individual-communal context with clear references to his African heritage. Equiano makes clear that his account is both personal and communal. While he begins with first person narration, he soon shifts back and forth between comments upon his character and the community he grew up in. Negotiating the private and the public, Equiano sets a pattern which will become a major paradigm for autobiographical writing of African, African American, and Caribbean writers in decades to follow. But his narrative reaches beyond a limited individual-communal connection, too. As he emphasizes, various communities may serve him as temporary sites to identify with. The inconsistency with regard to his cultural affiliations allows the narrator to create a flexible identity that enables him to connect and disconnect fairly quickly, as he is forced to move from nation to nation, culture to culture. What remains is a sense of play and ambiguity, the result of a writing process that translates the author's complex roles in the times of colonization into a multi-layered text. What emerges as a result of this mode is a hybrid text insofar as it is conceived and written within as well as against the terms of the dominant culture. Equiano is caught within the colonial discourse at the same time that he resists it. And, most important for my argumentation in relation to his "aesthetic of cool," Equiano presents himself consistently as a self in control over various textual discourses.

Thompson anchors "the aesthetic of the cool" in African and Black diasporic cultural life and production and defines it as "the sense of a deeply and complexly motivated, consciously artistic, interweaving of elements serious and pleasurable, of responsibility and of play" (1973, 41). As he explains, "[t]he telling point is that the 'mask' of coolness is worn not only in time of stress but also of pleasure, in fields of expressive performance and the dance. Control, stability, and composure under the African rubric of the cool seem to constitute elements of an all-embracing aesthetic attitude" (41). The aesthetic of the cool, according to Thompson takes center stage in thinking, acting, and creating. It suggests "composure, self-control, and social equilibrium" (41). As I read Equiano's autobiography, the narrating self puts on a mask of coolness providing a playful masquerade to move with dignity, agency, and sovereignty even in most difficult situations and confrontational encounters. And the aesthetic of the cool cherishes a creative and "aesthetic elaboration" (42). This explains why Equiano's writing makes us so often forget that *The Interesting Narrative of the Life of Olaudah Equiano* was also written as an abolitionist treatise. The author's use of humor, wit, double entendre, and, finally, his performative acts of creating a protean self express life experience in a positivist way despite the experience of enslavement.

The playfulness foregrounded by the frontispiece and the title is continued through Equiano's shifting narrative voices. Commenting upon the rhetorical devices Equiano resorts to, Henry L. Gates explains in his introduction to *The Classic Slave Narratives*:

> Equiano was an impressively self-conscious writer, and developed two rhetorical strategies that would be used extensively in the nineteenth-century slave narratives: the trope of chiasmus . . . and the use of two distinct voices. These two voices were meant to distinguish, in language, the simple wonder with which the young Equiano approached the world of his captor, and the more sophisticated vision, captured in a more eloquently articulated voice, of the author's narrative "present." The interplay of these two voices is as striking as Equiano's overarching reversal-plot pattern, within which all sorts of embedded reversal tales occur. Rarely would a slave narrator match Equiano's mastery of self-representation. (2002, 8–9)

Gates points out the presence of two different voices. And yet, I think, a close reading of Equiano's text reveals further differentiation in the author's rendering of voice. It is precisely on this level that creativity meets experience and shapes experience profoundly. As Vlad Pedre Glaveanu and others express in their recent "Advancing Creativity Theory and Research: A Socio-cultural Manifesto," "[c]reative outcomes are not only new and appropriate for a certain task; they can give meaning and even joy to our existence and, as such, represent a key marker of our humanity" (Glaveanu et al. 2019, 2). The shifts of voice correspond to the various roles Equiano takes on throughout the narration. Not only does he describe his roles as sailor, servant, and parson, the narrating voice itself becomes hybrid in that it speaks for Equiano as ethnographer, Equiano as tourist, and Equiano as political activist. At the same time, these creative strategies on the author's part fill the facets of self with life and express individual experience at the base of social, political, and cultural reflection. Clearly, these differentiations occur on what Gates labels "the more sophisticated" level of the narrative. And they demonstrate that hybridity is renegotiated in Equiano's autobiography.

While the phenomenon is generally considered a result of intercultural encounters, it is not a fixed form of cultural fusion. In times of slavery, in particular, hybrid identities become an essential survival strategy. Understanding, interpreting, and adopting practices of the colonizer are means to survive and a first step toward resistance, as Equiano's powerful use of the master's language emphasizes. Equiano, moreover, draws on hybrid identity formation since it allows him to transcend essentialist notions of being African. As a world traveler, he not only conquers space; he also gains insight into different economic and cultural practices to compare and borrow from. Moreover, we not only see him as a hardworking slave but also as an enterprising, at times even heroic character. As Equiano is eager to display by emphasizing a protean self, hybridity represents a fluid concept constantly changing within the dialectics of being enslaved and free, between necessity and choice. Hybridization in Equiano's rendering describes a process of growing into a fluid cosmopolitan identity, both expressing an early form of black cosmopolitanism (Nwankwo 2014) and fusing his own experience with the experience of power by the colonizer.[2] Klaus Benesch rightly classifies Equiano's memoir as "a hybrid, border-crossing text that continuously valorizes the colonized from within an ideology of the colonizer" (2003, 41).

As Benesch's statement emphasizes, Equiano clearly transgresses dividing lines between colonizer and colonized, master and slave. This means that neither his identity as a slave nor his identity as a free man is conceived as static. At first, he even introduces himself with a negative definition when he states, "I own offer here the history of neither a saint, a hero, nor a tyrant" (Equiano 2001, 19). Such a statement bears an anti-essentialist tendency and rouses the reader's curiosity about the author's identity. It further reveals that it is experience that is at stake here. Strikingly, identity is not primarily defined through 'race' in Equiano's text. Blackness itself is considered a hybrid phenomenon. Equiano writes the narrative where 'black' is generally caught between images of the ignoble and the noble savage. He playfully makes the reader aware that blackness may signify not only much greater multiplicity – "a multitude of black people of every description" (39) – but also economic and cultural equality in a transcultural context.

The way Equiano presents himself in the autobiography is consistent with the hybrid textual matrix. The autobiographical self stages a protean identity. The reader, accordingly, encounters various facets of Equiano's character giving voice to important life experiences in different contexts. It is important to note that the roles and masks Equiano takes on during the narrative are different and that their difference depends very much on the social and political status at any given moment. Equiano's displacements, in general, are imposed from the outside, since his status as enslaved person and servant hardly leaves space for individual decision. Accordingly, under these circumstances, most of his role playing acts

2 To underscore the mastery gained throughout the process of negotiating cultural differences, Equiano finally resorts to hybrid identities as a way to present himself as a playful picaro, for instance when he acts the parson in Georgia.

are performed out of necessity. At the same time, there are many moments in which it becomes clear that Equiano enjoys and plays with assuming various masks. Although Equiano makes clear that he, as a black writing subject, frequently confronts marginalization by race, nation, and class throughout the narrative, he at times crosses the borderline between margin and mainstream. He focuses on the marginalized status in his depictions of the middle passage and slave practices in the West Indies in particular. And yet, as a free man, he even assumes roles that are inextricably connected to practices of colonial power, and he becomes an agent within. Having acquired his freedom, he also becomes commissary for the Sierra colony and a temporary overseer of Dr. Irving's plantation on the Muskito coast.

A closer look at the historical context reveals why other factors such as class, social status, and cultural background play a more significant part in the identity formation process than race does in Equiano's autobiography. Eighteenth-century slave trade was central to the building of colonial empires. And it was primarily based upon economic rather than racial considerations. Moreover, as Paul Baepler's collection of barbary captivity narratives illustrates in the commentary on enslavement of Europeans by North Africans, slavery did not respect racial boundaries (Baepler 1999). Both, African as well as white masters, were aware of the economic profit that slavery provided them with. Equiano, too, reports the enslavement of whites. He refers to the white galley slaves in Italy and points out that Turks enslaved Greek workers. Although he comments upon racial issues supporting his abolitionist cause, he foregrounds the economic matrix for the expansion of the slave trade. And, as Equiano's success story illustrates, economic power and social status mark his entry into the British society.

Equiano further dismantles the significance of race by playing with different ethnic affiliations: "[D]id I consider myself an European, I might say my sufferings were great: but when I compare my lot with that of most of my countrymen, I regard myself as a particular favourite of Heaven" (2001,19).

Clearly, his claim for a European identity remains hypothetical in the above context. Still, Equiano signals that he may consider such an identity for himself in spite of racial differences. Both, as traveler and "travelee," Equiano absorbs numerous cultural influences which shape his experiences of self and other throughout the narrative.[3] His shifting position again underscores that he does not express the views of the colonized only, as the critical link of his text to the genre of the slave

3 Here I am using the terminology developed by Geraldine Murphy. She pursues both dimensions of travelling. To her, "Equiano, the travelee ripped from his family and culture and westernized perforce, becomes a traveller himself when he grows up, observing Europe and other continents just as Europeans had observed Africa" (2001, 369).

narrative may suggest. Equiano adopts the colonizer's gaze, too, meaning that, in spite of its abolitionist orientation, his text refuses to be simply classified as a narrative of subversion and resistance. As William L. Andrews puts it, "[t]elling freedom can involve demonstrating one's liberation from any outworn or restrictive form of thinking or writing" (1989, 90). Equiano "plays it cool," letting his own experience and his reflections of it in various contexts become the key for his thoughts on slavery, history, and coexistence.

'Cosmic Zora' and performativity of self in Hurston's "How It Feels to Be Colored Me"

When it appeared in 1928, Zora Neale Hurston's autobiographical essay meant a provocation to large parts of the black and white readership. Unabashedly, she based her flamboyant, yet controlled celebration of black female self-empowerment on her own experience of joy, pleasure, and achievement. According to Mariel Rodney:

> After a spectacular debut on the Harlem literary scene barely three years prior, Hurston's 1928 essay "How It Feels to Be Colored Me" dramatized an unexpected portrait of the racial artist for the unwitting readers of the *World Tomorrow*, an American political magazine that catered to a white readership composed mostly of women. The essay positioned Hurston as a new kind of public intellectual and an artist keen on articulating the intersecting stakes of publicity and spectacle, on the one hand, and quotidian performances of interiority, race, and racialization, on the other hand. (2020, n. pag.)

Next to her autobiography *Dust Tracks on the Road* (1942) and her recently posthumously published biographical work *Barracoon: The Story of the 'Last Black Cargo'* (2018), Hurston's essay is among her most remarkable contributions to a larger tradition of life writing, which has become a canonical text for reflections on race and identity. As anthropologist and folklorist, Hurston demonstrated her deep interest in what Melville Herskovits calls 'Africanisms' or "traces of African Heritage in the customs of 'New World Negroes'" (2018, 1). In Raquel Kennon's account, "On 14 December 1927, Zora Neale Hurston began a three-month period of ethnographic fieldwork and folklore collection by revisiting octogenarian Oluale Kossola in Africatown, also called Plateau, a postbellum independent society established and controlled by Africans, located a few miles north of Mobile, Alabama" (2021, 5). Hurston's desire to listen to and record and transcribe Kossola's story shows that she wanted to fill a gap in the historical archive of the transatlantic slave trade. In *Barracoon* she reveals her motivation: "but not one word from the sold. The Kings and Captains whose words moved ships. But not one word from the cargo" (2018, 6).

While her concern with the past characterizes her anthropological work in *Barracoon* and her folklorist work in anthologies like *Mules and Men* (1935), Hurston takes a seemingly surprising stance in her autobiographically framed essay "How It Feels to Be Colored Me," in which she consciously downplays colonial enslavement and dismisses the importance of this memory for constituting her own life tracks as a colored woman in the twentieth century: "Slavery is sixty years in the past. The operation was successful and the patient is doing well, thank you" (2020, 528), she boldly responds to those who remind her of past enslavement of African Americans. Responses like these show her in an imaginary conversation with others and underscore the performativity of her writing in this short autobiographical essay. She chooses a minimalist version of the essay genre and proceeds in an episodic way in which important periods and moments in her life are compressed in a time-space continuum allowing her reflections about her self- and identity-making in the process. Mostly growing up in the all-black town of Eatonville, Florida, which now holds the Zora Neale Hurston Museum of Fine Arts and annually stages the Zora Neale Hurston Festival of the Arts and Humanities since 2004, Hurston's childhood is framed by solid African American communal and institutional life. She presents her background as a backdrop against which she projects a cheerful, extroverted, and self-reflective self of a young girl growing up. Writing this autobiographical piece at the time when she had taken center-stage in the literary salons during the heydays of the Harlem Renaissance, she vividly drew upon her presence in the public sphere and combined experiences from her childhood with her public presence in New York. There is a strong awareness of self and how the self is perceived by a larger public when she admits: "It is quite exciting to hold the center of the national stage, with the spectators not knowing whether to laugh or to weep" (528). This statement coincides with an earlier observation about childhood times in Eatonville in which she reveals her extroverted character: "The front porch might seem a daring place for the rest of the town, but it was a gallery seat to me. My favorite place was atop the gatepost" (527). In passages like these, her essay reads like a soliloquy and the spatial references stress the theatrical performativity underlying Hurston's presentation of self. As Glaveanu and others stress in their manifesto, "creative action is at all times relational" (2019, 2). While zooming in on her own self, Hurston always seems to be aware of the other she addresses directly or indirectly. "There is no form of human creativity that does not rely on direct, mediated or implicit social interaction or exchanges," the authors of the manifesto conclude (2).

The most powerful creative element in Hurston's essay lies in her exploration of performativity of body, voice, and reflected self as means of expressing individualized black experience in reference to a dissident black performance culture. According to Daphne Brooks, African American women performers like the Hyers Sisters already in late nineteenth century "employ[ed] their own bodies as canvasses of dissent in popular performance culture" (2006, 6). She stresses "the skill of the performer who, through gestures and speech as well as material props and visual technologies, is able to confound and disrupt conventional constructions of the racialized and gendered body . . ." (8). The front porch becomes Hurston's first important stage from where she sings and dances to and for the tourists and locals passing by. Her choice of the "gatepost" as "gallery seat" (Hurston 2020, 527) shows that she is a performer and a voyeur at the same time. While her body dances and her voice sings, she is attentively watching her audience. It is her voyeuristic gaze that meets the visitors and that actively propels the interaction between her and them; a form of social interaction that becomes a profitable economic experience since the tourists tip her generously. Referring to the porch, Mariel Rodney maintains that this is the site "where [Hurston] will practice negotiating a gendered, racial gaze, while leveraging that space as one of pleasure and power" (2020, n. pag). In a letter to Hurston, the leading Harlem Renaissance figure Alain Locke criticized her emphasis on performance and minstrel calling her essay "a mistake" (2012, n. pag.) evoking times of black-face minstrelsy and coloniality. As this letter confirms, Hurston's performative twist was not agreeable to racial uplift intellectuals like Locke who were eager on creating a new respectability of black culture. But Hurston's essay is very much autobiographical in that within it experience meets creativity. Experience for Hurston includes joy and pleasure that she finds in her own performances as well as in the musical event she attends with a white friend in New York. For Hurston, joy and pleasure are essential human experiences that unite the emotional with the cognitive and they are empowering forces that strengthen the self in dialogue with her audience.

With a nod to Thompson's discourse of 'an aesthetic of the cool' – "composure, self-control, and social equilibrium" (1973, 41) – I would like to argue that Hurston demonstrates precisely this when she enters into an imaginative conversation with the predominantly white female readership of the *World Tomorrow* in 1928. She shows no intentions either to uplift the race or to see herself as inferior to white folks. Rather, her creative play with words is a sheer act of self-empowerment: "It is clear that I was the first 'welcome-to-our-state' Floridian" (Hurston 2020, 527). Repeatedly crossing the lines between oral and written discourse thus mixing African and African American modes of autobiographical writing, she creates a performance-oriented narrative including elements of speech, elocution, and interior monologue. In all this, the imagined audience is nothing less than the nation itself. She uses the nation's rhetoric of myth-making and turns it into her own: "No one on earth had ever a greater chance for glory."

As she jokes, "The position of my white neighbor is much more difficult" (528). While she exposes that she is both object and subject within her own discourse, "[s]ometimes, I feel discriminated against, but it doesn't make me angry" (529), she always responds with a corrective, at times provocative response: "How can any deny themselves the pleasure of my company!" (529). Despite its celebratory, at times flamboyant nature, Hurston's autobiographical essay is very much concerned with returning to and establishing a self-reflective equilibrium. This is especially true when Hurston becomes her own audience when she stands back and evaluates her own discourse: "But in the main, I feel like a brown bag of miscellany propped against a wall. Against a wall in company with other bags, white, red and yellow" (529).

Hurston's "proclivity for contrarianism" (Kennon 2021, 90) is well established. As Kennon observes, Hurston's style is one of "provocateur, probing, unsettling orthodoxies, revealing contradictions, and complicating white-black racial histories" (90). Similarly, Tiffany Ruby Patterson explains that "it is precisely Hurston as faithful, unsparing 'chronicle[r] of black life' that has been used to denounce her as a traitor to the race" (2005, 5). Against this backdrop, we need to see Hurston's writing in the context of her double function as anthropologist, trained by Boas among others, and creative writer of fiction. What she does in "Colored Me" is to look at her own self through anthropological eyes. She is self and other. This gives her the privilege to occupy a double subject and object position. Likewise, it gives her license to bridge the gap between biography and autobiography, between anthropological narrative and personal life story. While Gates is referring to her "double-voiced" attitude in many of her writings, I search for a double agency in "Colored Me" that makes her performance of self all the more powerful. It is in the moving back and forth between writing fiction and writing anthropology that Hurston unfolds her full potential for a creativity that defines the self and redefines the social. "She celebrated herself through words, crafted herself, affirmed herself, and perhaps most important and most primal, named herself" (Holloway 1987, 115). Consciously or unconsciously, but certainly with emphasis, Hurston performs the "aesthetic of the cool" in her autobiographical/anthropological self-definition in "How It Feels to Be Colored Me." She is continuously dedicated to creating an equilibrium between self and world without obliterating difference. The narrative progression puts her 'self' in various spatial and situational contexts. Meandering between performative self, anthropological gaze, and writerly introspection, she describes her 'self' as both being and becoming. What unfolds is therefore a rooted self with flexibility to adapt, resist, and control situations. It is thus accurate to say that her understanding of identity and history goes beyond binary opposition and easy racial divides. As Rodney puts it, "Hurston uses a performative and syntactical economy of 'color' to stage the vibrancy and the opacity of Blackness in ways that depart from her contemporaries" (2020, n. pag). Hurston shows us that experience is at the base of Blackness and the latter may vary from context to context. The fact that she writes about "how it feels" highlights that "color" in Hurston's rendering becomes a complex and fluid set of emotional, cognitive, and performative components. "I am not tragically colored . . . . No, I do not weep at the world – I am too busy sharpening my oyster knife" (2020, 527). Employing the autobiographical mode as spectacle and exposing her own pleasure in performing Blackness, she lets creativity speak and express subjective experience, thus deconstructing the prevailing discourses of race at the time.

Conclusion

In conclusion, let me stress that Equiano and Hurston embrace life writing as creative laboratory for performative self-expression, challenging prevalent cultural and racial discourses. Their embrace of experience and individuality as valid sources for literary and anthropological expression provides them with a subjective control over authorship and agency. As I have shown throughout this essay, each in their own way practice life writing as personal and social creativity and celebrate a sense of empowerment, balanced and expressed through an aesthetic of the cool that goes far beyond claiming subject status, citizenship, and belonging: While Equiano's colonial text becomes a playful and sophisticated argumentation against any justification of slavery, Hurston's text creates a twentieth-century vanguard black female voice that challenges the public discourse as it celebrates relational differences beyond questions of race and gender.

Works cited

Andrews, William L. 1989. "Toward a Poetics of Afro-American Autobiography." In *Afro-American Literary Study in the 1990s*, ed. Houston A.Baker, Jr. and Patricia Redmond, 78–9. Chicago: University of Chicago Press.

Alexander, Elizabeth. 2019. "The Genius of Romare Bearden." In *The Romare Bearden Reader*, ed. Robert O'Meally, 185–19. Duke University Press.

Baepler, Paul, ed. 1999. *White Slaves, African Masters: An Anthology of American Barbary Captivity Narratives*. Chicago: University of Chicago Press.

Banerjee, Mita. 2018. *Medical Humanities in American Studies: Life Writing, Narrative Medicine, and the Power of Autobiography*. Heidelberg: Winter.

Benesch, Klaus. 2003. "The Pitfalls of Memory: Cross-Cultural Discourse and Autobiographical Form in Olaudah Equiano's Interesting Narrative." In *Sites of Memory in American Literatures and Cultures*, ed. Udo Hebel, 31–42. Heidelberg: Universitätsverlag Winter.

Brooks, Daphne. 2006. *Bodies in Dissent: Spectacular Performances of Race and Freedom 1850–1910.* Durham: Duke University Press.
Brown, Lisa R. 2011. "Caribbean Life-Writing and Performative Liberation: Beyond the Boundaries." In *The Routledge Companion to Anglophone Caribbean Literature*, ed. Michael A. Bucknor and Alison Donnell, 276–84. London: Routledge.
Butler, Judith, and Athena Athanasiou. 2013. *Dispossession: The Performative in the Political.* Cambridge: Polity Press.
Constanzo, Angelo. 2002. "From the Spiritual Autobiography and Slave Narrative of Olaudah Equiano." In *The Norton Critical Edition of The Interesting Narrative*, ed. Werner Sollors, 348–50. New York: Norton.
Egan, Susanna, and Gabriele Helms. 2004. "Life Writing." In *The Cambridge Companion to Canadian Literature*, ed. Eva-Marie Kröller, 216–40. Cambridge: Cambridge University Press.
Equiano, Olaudah. 2001. *The Interesting Narrative of Olaudah Equiano or Gustavus Vassa, the African, Written by Himself*, ed. Werner Sollors. New York: Norton.
Equiano, Olaudah. 2002. "Introduction." In *Classic Slave Narratives*, ed. Henry L. Gates, Jr, 1–14. New York: Signet.
Glaveanu, Vlad Petre, and Michael H. Hanson, 2019. "Advancing Creativity Theory and Research: A Socio-Cultural Manifesto." *The Journal of Creative Behavior* 54, no. 3 (January): 1–5.
Holloway, Karla F.C. 1987. *The Character of the Word: The Texts of Zora Neale Hurston.* Westport: Greenwood Press.
Hornung, Alfred. 2009a. *Autobiography and Mediation.* Heidelberg: Winter.
Hornung, Alfred. 2009b. "Transcultural Life-Writing." In *The Cambridge History of Canadian Literature*, ed. Coral Ann Howells and Eva-Marie Kroeller, 536–55. Cambridge: Cambridge University Press.
Hornung, Alfred, ed. 2013. *American Lives.* Heidelberg: Winter.
Hurston, Zora Neale. 2020. "How It Feels to Be Colored Me." In *The Glorious American Essay: One Hundred Essays from Colonial Times to the Present*, ed. Phillip Lopate, 526–30. New York: Pantheon Books.
Hurston, Zora Neale. 2018. *Barracoon: The Story of the Last "Black Cargo."* Ed. Deborah G. Plant. New York: Amistad.
Kadar, Marlene. 1992. *Essays on Life-Writing.* Toronto: University of Toronto Press.
Kennon, Raquel. 2021. "'In de Affica Soil': Slavery, Ethnography, and Recovery in Zora Neale Hurston's *Barracoon: The Story of the 'Last Black Cargo.'*" *Melus* 46, no.1 (Spring): 75–104.
Laforest, Marie H. 1996. "Black Cultures in Difference." In *The Post-Colonial Question: Common Skies, Divided* Horizon, ed. Iain Chambers, and Lidia Curti, 115–20. London: Routledge.
Locke, Alain. 2012. Letter to Zora Neale Hurston, 2 June 1928, quoted in Eric Watts, *Hearing the Hurt: Rhetoric, Aesthetics, and the Politics of the New Negro Movement.* Tuscaloosa: University of Alabama.
Maftei, Micaela. 2013. *The Fiction of Autobiography.* New York, NY: Bloomsbury.
Moore-Gilbert, Bart. 2009. *Postcolonial Life-Writing: Culture, Politics, and Self-Representation.* New York: Routledge.
Murphy, Geraldine. 2001. "Olaudah Equiano, Accidental Tourist." In *The Interesting Narrative of Olaudah Equiano or Gustavus Vassa, the African, Written by* Himself, ed. Werner Sollors, 368–82. New York: Norton.
Nwankwo, Ifeoma Kiddoe. 2014. *Black Cosmopolitanism: Racial Consciousness and Transnational Identity in the Nineteenth-Century Americas.* Philadelphia: University of Pennsylvania Press.
Patterson, Tiffany Ruby. 2005. *Zora Neale Hurston and a History of Southern Life.* Cambridge: Harvard University Press.

Rodney, Mariel. 2020. “Staging Black Effects: Hurston’s How It Feels to Be Colored Me.” *S&F* online edition, February 2. Guest edited by Monica L. Miller and Tami Navarro. https://sfonline.barnard.edu/zora-neale-hurston/staging-black-affects/.

Smith, Sidonie, and Julia Watson. 2010. *Reading Autobiography: A Guide for Interpreting Life Narratives*. Michigan: University of Minnesota Press.

Sollors, Werner, ed. 2001. “Introduction.” In *The Interesting Narrative of Olaudah Equiano or Gustavus Vassa, the African, Written by Himself*, ix–xxxi. New York: Norton.

Thompson, Robert Farris. 1973. “An Aesthetic of the Cool.” *African Arts* 7, no. 1 (Autumn): 40–91.

Walder, Dennis. 2011. *Postcolonial Nostalgias: Writing, Representation, and Memory.* New York: Routledge.

Astrid Haas

Black Narrative Self-Making in a Changing Scene: The *Life and Adventures* of James Beckwourth as Western Borderlands Autobiography

Introduction

The present essay analyzes James Beckwourth's narrative self-making through the ways he portrays his life in a changing North American West in his autobiography, *The Life and Adventures of James P. Beckwourth* (1856). During the first half of the nineteenth century, this region underwent profound social and environmental changes, which nurtured a vivid public interest among Americans in the landscapes and lifeways of the West. Western "frontier" or borderlands autobiography was one of the media formats that capitalized on this curiosity and simultaneously provided Westerners – and others who pretended to have lived in the West – with opportunities to contribute to U.S. print culture. Within this body of writing, *The Life and Adventures of James P. Beckwourth* enjoys a unique place. During his life, Beckwourth adopted different identities – most likely to obtain personal opportunities, enjoy a certain degree of independence, and live a life of independence and adventure. Working as a trapper, trader, scout, and Crow warrior of Anglo and African descent, he moved across ethnic identities and professional occupations in the Great Plains, the Rocky Mountains, and California. His *Life and Adventures* is one of few works of African American autobiography testifying to Black experiences in the region from 1800–1850. Drawing on autobiography research, Black (and) Indigenous, New Western and Border studies, my research argues that, and shows how, Beckwourth creatively appropriates the genre of the borderlands autobiography to obtain social agency and authority as a Black man in a transforming region. I specifically analyze the role his different identities played in his narrative self-making as a competent professional, and how this versatile self-fashioning rendered Beckwourth a problematic Western "hero" who defied ethnocultural and political boundaries.

https://doi.org/10.1515/9783111551968-006

The nineteenth-century North American West

The nineteenth-century North American West – the region west of the U.S. western border that was annexed to the country between 1783 and 1853 – constituted what scholars define as a borderland region. In contrast to the politically fraught concept of the American "frontier," the notion of the borderland does not imply a westward-moving demarcation line between white "civilization" and a natural and Indigenous social "wilderness." Instead, it refers to a contested contact zone of shifting spatial and cultural boundaries (Adelman and Aaron 1999, 815; LeMenager 2003, 684) where Native and colonizing populations "competed for resources and land but also shared skills, foods, fashions, customs, languages, and beliefs" (Hämäläinen 2008, 7). While Indigenous nations had earlier encountered Spanish soldiers and missionaries or French and British traders, this borderland underwent a more profound political, economic, social, and ecological transformation during the nineteenth century. Between 1800 and 1850, peace treaties with Native nations, land acquisitions like the Louisiana Purchase (1803), or military gains from the U.S.–Mexican War (1845–1848) formally expanded the United States' territory across the West (Carlson 1998, 124–41; Nugent 2008). Continuing practices from the colonial period, the Anglo-American commercial and agrarian appropriation of the land through trade and white settlement, accompanied by Indigenous removal and genocide as well as the oppression of Mexicans, gradually brought this territory under de-facto United States control (Doolen 2014, 6, 8–10; LeMenager 2003, 684–85, 691). Moreover, environmental factors like predation, fire, or climate change, as well as the impact of hunting, trade, white migration, land cultivation, and livestock-raising, further impacted Western nature, reducing regional wildlife and plant diversity (Carlson 1998, 136; Isenberg 2004, 80–87).

During the nineteenth century, the Western borderlands also provided growing prospects to African Americans, a phenomenon that has been studied in both Black Western (Ravage 1997; Taylor 1998) and Black Indigenous (Mays 2021; Roberts 2021) Studies, in the latter, though with a focus on the post-Civil War era. During the first half of the century, enslaved Blacks were forced to accompany their white or Indigenous owners who moved – voluntarily or through enforced removals – to the region (Haas 2022, 257; Roberts 2021, 12–14, 23–24). In contrast, free, manumitted, and self-emancipated African Americans began to flock westward seeking economic improvement, freedom, and/or an escape from racialized social restraints (Ravage 1997, 9–10, 47–48; Taylor 1998, 48, 81–82). Among whites, many Black newcomers either set up their own businesses or found only menial jobs (Ravage 1997, 29–36, 39–44, 51–56). They fared better in Mexican territories before their annexation to the United States in 1848, as Mexico abolished slavery in 1829 and granted Black people social rights (Taylor 1998, 37–42, 53–54). In other parts of the West,

African Americans enjoyed greater liberties in the Indigenous borderlands – the territories home to the region's original nations. They particularly succeeded in professions that brought them into contact with Native populations, such as fur trappers, traders, or negotiators accompanying scientific expeditions among the Plains Indian nations (Browder 2000, 117; Lape 2000, 30–31). Many of them also forged close ties with Plains bands, adopted their customs, and/or took Indigenous wives (Hoxie 1995, 57, 70–71; Taylor 1998, 48).

Western Borderlands autobiography and *The Life and Adventures of James P. Beckwourth*

As white explorers, traders, soldiers, and settlers appropriated the West, the region aroused a growing public interest among audiences in the Eastern United States, who viewed the unfamiliar landscapes and lifeways of the mountains and plains with fascination and cultural anxiety (LeMenager 2003, 686, 691–92, 696–97). Among many other works of writing, ranging from scientific expedition accounts to popular fiction, a new type of autobiography successfully capitalized on this curiosity. The Western "frontier" or borderlands autobiography was commonly authored by prominent non-Indigenous people, primarily Anglo-American men, whose activities in the nineteenth-century West helped prepare the region for white settlement through violently subjugating resident Native populations. These texts tell Anglocentric stories of "heroic" individuals overcoming the dangers of untamed nature and antagonistic populations. In contrast to life narratives from the Eastern United States, Western autobiography also shares major features with the Indigenous autobiography of the time: As many people in the borderlands had limited literacy, they often dictated their stories to more formally educated white amanuenses. Their texts blend elements of Western oral and Eastern American written narrative traditions such as the picaresque novel, Indian captivity narrative, frontier romance, and Western "tall tale" of exaggerated exploits of fictional or historical folk heroes from the region. Moreover, both Western and Indigenous autobiography fashion their usually male protagonists as "men of action" rather than lettered "men of culture" to validate their experiences and authority to speak to white, educated audiences. In so doing, these texts not only provide readers with first-hand knowledge about unfamiliar places and people but also cater to audiences' desire for adventure, heroism, and shrewdness. While they convey a fascination with cultural and environmental alterity, non-Indigenous borderlands autobiographies simultaneously profess loyalty to Anglo-American civilization to tone down Eastern readers' anxieties about a region where ethnic and social boundaries were frequently blurred. They further legitimize the genocidal removal of Western Native nations to appropriate their lands for white settlement as "the inevitable triumph of 'civilization' over 'savagery'" (Krupat 1981a, 314; also 309–13; Krupat 1981b, 33–34; Lape 2000, 23, 27, 36).

With its blend of narrative genres and emphasis on acts of self-making, Western autobiography renders visible the performative nature of the genre. I follow Sidonie Smith and Julia Watson in understanding

> autobiographical occasions as dynamic sites for the performance of identities that become constitutive of subjectivity. In this view, identities are not fixed or essentialized attributes of autobiographical subjects; rather, they are enacted and reiterated through cultural norms and discourses, and thus remain provisional and unstable. (2010, 214)

It is thus the process of narrating one's life story that performatively constitutes the self a work of autobiography claims to merely portray (214). This performative self-making through autobiographical narrative can be observed particularly well in one of the most intriguing works of Western borderlands autobiography. James Beckwourth's *The Life and Adventures of James P. Beckwourth* is one of only a few black-authored examples of the genre, as African Americans only made up between 1.5 and 3 percent of the Western population in the nineteenth century (Ravage 1997, xix) and were even less represented in nineteenth-century U.S. print culture. White-authored texts of the time either ignore Black Western life or reiterate the prevailing stereotype of a populace "unable to speak well, reason intelligently, or conduct themselves with socially acceptable manners or mores" (Ravage 1997, 27; also, 26–29). In a context in which "the concept of conquest dominated American attitudes towards nature," they contributed to a white public discourse that framed Blacks as beast-like toilers of the land. This discourse justified Black economic exploitation and cultural exclusion from the "dominant narrative of mastering nature" (Gerhardt 2002, 522; also, 520) central to non-Indigenous Western identity and autobiographical self-fashioning.

Against this backdrop, James Beckwourth (1798/1800–1866) claimed agency and authority not only as a Black "frontiersman" in the Western borderlands but also as an African American and Crow autobiographer in white-dominated U.S. print culture. He had been born into slavery in Virginia but relocated to Missouri with his slaveowner father's household when he was still a child. Manumitted as a young man, he joined the Rocky Mountain Fur Company in 1824 to explore Indigenous territories in the Northern Plains and the Rocky Mountains, trap beaver, and trade with the resident Native nations. Two years later, Beckwourth joined a band of the Crow nation. Rising as a warrior among their ranks, he lived among them for thirteen years. He subsequently worked as a trader and innkeeper in various places in the Rocky Mountains, Southern Plains, and California but also as a messenger and scout for the U.S. Army in the Seminole Wars in Florida, the U.S.–Mexican War, and against different Plains Indian nations in 1864 (Oswald 1981, 603–28; Wilson 1988).

In 1856, *The Life and Adventures of James P. Beckwourth* appeared with the eminent publisher Harper's in New York. Although he was literate, Beckwourth had hired Thomas Bonner, a white confidence man, journalist, and justice of the peace, to pen his autobiography. The resulting narrative is a collaborative work, which shows in its speaking with a monological narrative voice but frequently mediating, if not ventriloquizing it (Couser 2004, 35–39, 48). While some linguistic mannerisms were typical of Beckwourth (Wilson 1988, 61), the elaborate language, literary quotes, European cultural references, and use of genre conventions indicate Bonner's hand (Lape 2000, 23–25; Madera 2015, 90–91) and thereby place the author's Black voice within a white-centric discourse. This also becomes manifest in the reception of the volume. *The Life and Adventures* met with public acclaim but also aroused controversy over Beckwourth's boasting, factual inaccuracies, and racial passing (Oswald 1981, vii–xi; Lape 2000, 25–27; Wilson 1988, 5–7, 196). However, despite the restrictions imposed on his voice, I follow Rafia Zafar's view that "the black self can materialize in a white frame, however incompletely" (1997, 55). As he actively selected his amanuensis and authorized Bonner's manuscript for publication (Oswald 1981, 604), Beckwourth retained some control over his publicized story that mitigated the social power difference between the Black author-narrator-subject and his white editor-scribe (Couser 2004 37, 41–44, 48). Moreover, Beckwourth was aware of the value his "career of wild adventure and thrilling romance" (Beckwourth 1981, 534) and knowledge of Plains Indian cultures had to white readers hungry for ethnographic information and captivating tales about the West. Analyzing his text as the Western autobiography of a Black/mixed-race frontiersman and member of the Crow nation thus provides new insights for Black Indigenous and autobiography studies. In the remainder of this essay, I will examine the strategies of narrative self-making in a changing scene Beckwourth employs in his book.

Narrative self-making through cultural affiliation

As a Western borderlands autobiographer, Beckwourth's authority rests on his narrative self-making as an eminent "frontiersman." He achieves this by fashioning himself as an honest intercultural mediator, heroic individual, and loyal servant of the United States' westward territorial expansion and transformation. To do this, he draws on his familiarity with the nature and Indigenous populations of the Western borderlands. In his *Life and Adventures*, he claims authority especially on the Plains nations, as he, "by living among them, ha[d] learned their language and hence become acquainted with the nature of their divisional lands" (Beckwourth 1981, 26). Such a connection between nature, culture, and the social was a staple of nineteenth-century United States discourse about North American nature and its Native societies. This discourse viewed land as a potential economic resource that permitted white settler communities to emerge and Euro-American civilization to spread in the West. Indigenous populations were seen as an integral part of nature rather than civilizations with claims to the territory, as they did not cultivate the soil for capitalist gain. By thus declaring the Indigenous nations unfit to own their land, this discourse prepared and justified their violent removal from large parts of the West (Doolen 2014, 154–55; Haas 2021, 6).

Beckwourth's *Life and Adventures* uses a variety of narrative strategies to present its narrator in his roles as an intercultural mediator, heroic adventurer, and American expansionist. The author had been among a number of non-Indigenous trappers and traders in the Western borderlands who joined the Crow nation during the 1820s when trade relations between the Crow and U.S. business companies intensified. For the Crow and many other Indigenous nations, tribal membership was based on biological or social kinship, which could include people of other origins who lived among a tribal band and adapted to its lifeways (Hoxie 1995, 43, 57, 65–66, 70–71). Beckwourth possibly joined the Crow as an envoy of the American Fur Company (Wilson 1988, 48–49, 60–61). However, in his autobiography, he presents his arrival among the Crow as a story of Indigenous capture and adoption (Beckwourth 1981, 145–50; Haas 2022, 261–65). The literary genre of the Indian captivity narrative resonated widely with American audiences, as Indigenous bondage had formed part of the white settler experience in North America since the seventeenth century and continued to occur in the West in Beckwourth's days. Also, various Native nations, including the Crow, were known to integrate captives into their bands (Haas 2022; Sayre 2000). Framing his entrance among the Crow as a captivity-cum-adoption story legitimized Beckwourth's subsequent life as an Indigenous warrior. This strategy thus "offered [the black author] an alibi for transculturation, an assurance against cultural regression, and thereby credibility with mainstream readers" (Sayre 2000, 14).

To further maintain this credibility and to claim a place among a "civilized" society that usually denied Black men, Beckwourth had to present himself as having adopted and simultaneously rejected Indigenous culture. He does this in his autobiography by fashioning himself as a valued Crow warrior precisely for his "superior" non-Indigenous intellect, character, and culture. For example, he foregrounds his Christianity by calling the Crow "a people really pagan" who "knew nothing about . . . our Savior" (Beckwourth 1981, 120, 162). He likewise demonstrates his entrepreneurialism and loyalty to American society when he states his "intention . . . to induce the Crow to devote their undivided attention to trapping, not alone for their own benefit, but [also] for the interest of the company in whose service I was engaged" (220). Later, he justifies his lucrative whisky trade among the Cheyenne despite being aware of how this "most infernal practice" (444) impacted Indigenous lives, as he emphasizes how he ensured that his Native trading partners would purchase more essential goods before receiving any alcohol (433–34, 440–48, 459–61).

Aligning with the dominant Anglo-American social discourse (Madera 2015, 94), the narrator further draws on the popular tropes of Indigenous people as dangerous "savages" and immature "children of the forest" (Beckwourth 1981, 161, 198) to legitimize their subjection to white American rule and civilization. He omits any reference to the intensifying armed conflicts among Native nations over contested territories and lucrative trade opportunities resulting to a large extent from intensified white commerce, land-taking, and Indigenous removal (Hoxie 1995, 55–56, 61, 65–67, 70–72, 78). Instead, he depicts the Crow in his narrative as a strategically inept yet belligerent nation that depended on his non-Indigenous bravery, diplomatic and tactical skills and whose "repeated acts of cruelty" repelled him (Beckwourth 1981, 198). Presenting himself as advancing the westward territorial expansion and Anglo-Americanization of the Western borderlands, Beckwourth's narrator specifically justifies his participating in the violent Crow warfare his audience would otherwise not have endorsed (Lape 2000, 29–30). "My body was among the Indians, but my mind was far away from them and their bloody deeds" (Beckwourth 1981, 120), he declares. He further claims to have "fought . . . against the most relentless enemies of the white man" as a Crow warrior and made sure that the Crow "never shed the blood of the white man during my stay in their camp" (232, 198). Similarly, he argues that he intended to induce fear among rivaling Indigenous nations to ensure that the Crow could trade in peace (361). On the grounds of these achievements, Beckwourth asserts his identification with white civilization as "an American citizen" and "sincere lover of my country" (529).

Beckwourth further establishes his loyalty to U.S. society through his portrayal of Mexicans in the chapters of his *Life and Adventures* depicting his life in New Mexico and California. In contrast to contemporary Anglo-American discourses on Mexicans (Haas 2021), Beckwourth's narrator bases his attitude toward this population not on their ethnic identity but on their responses to the United States' territorial expansion and political rule in their native regions. While he praises the determination and bravery of the Mexican inhabitants of California who fought against Mexico in the U.S.–Mexican War (Beckwourth 1981, 466–67), he scrutinizes the Hispanic New Mexicans for the Taos Rebellion/Revolt, an 1847 uprising of local Mexican and

Pueblo Indigenous residents against the Anglo-American occupation of their town. Tellingly, in describing the impact of this insurgency, the narrator recurs to cultural and religious imagery that frames this uprising in moral rather than, more accurately, political terms. As he compares the Taos Rebellion to the 1572 St. Bartholomew's Day massacre of Huguenots by Catholic mobs in France, he appeals to his American target readers' anti-Catholicism. Similarly, in asserting that "the barbarities we witnessed [in Taos] exceeded in brutality all my previous experience with the Indians" (185), the narrator implies that the Mexicans were even less "civilized" than the Indigenous nations whose violence he had recounted earlier. He once again asserts his own belonging to the United States when he speaks of the rebellion's Anglo-American victims as "our murdered fellow-countrymen" (485).

Despite thus affirming his Americanness, Beckwourth's description of his life among the Crow in his *Life and Adventures* also validates Indigenous civilization, even though it is marred by the narrator's stereotyping of Native nations as wild offspring of nature. In nostalgic reverie, he admits his attachment to the Crow, stating that "my heart turns naturally to my adopted people" (Beckwourth 1981, 535). Elsewhere he praises their bravery, communalism, loyalty, and the liberty he enjoyed among them (161, 383–84). Beckwourth's book further blurs the boundaries between Indigenous and non-Indigenous populations in the Western borderlands by pointing out certain similarities they shared. For instance, the narrator repeatedly mentions white settlers, explorers, and trappers engaging in practices popularly associated exclusively with Native nations, such as scalp-taking or invoking the "Great Spirit" (16, 78, 104). "If the Indian was unrelenting, and murdered with his lance, his battle-axe, and his knife, his white brother was equally unfeeling, and had ways of torturing his victim" (383), he further emphasizes. Beckwourth even confirms Indigenous land rights through an analogy with countries like the U.S.A. rejecting foreign interventions on their territories: He states that the Crow "resent in their hearts the invasion of the immigrant just as any civilized people would, if another nation, without their permission, should cross their territory" (528). As he foregrounds the parallels and connections among Indigenous and non-Indigenous people in the West, Beckwourth undermines the social construction and naturalization of the prevailing ethnoracial hierarchy among different populations. In so doing, he legitimizes his mobility across cultural and geographic boundaries as a valid path to obtain social agency and economic opportunities otherwise unavailable to him.

Resisting boundaries through racial passing

Beckwourth crosses and blurs socio-spatial boundaries in his *Life and Adventures*, especially through his racial passing and cultural appropriation. In his text, he fashions himself as a white man who adopts a Crow Indigenous identity. United States social discourse does not view racial passing as a legitimate move to escape social constraints. Instead, it frames passing as an illegitimate act of treason and fraud – denying a seemingly unchangeable biological identity obtained by birth such as race to obtain another, usually more privileged, one (Ginsberg 1996, 1–5, 8, 13). For his white contemporaries who knew him to be Black or mixed-race, Beckwourth's passing for white rendered him an illegitimate "ethnic impersonator" (Browder 2000, 2). Typical of many such impersonators (112), Beckwourth's racial passing did not aim to challenge the structural discrimination of either African Americans or Indigenous nations but represents the upward social movement of an individual seeking opportunities otherwise inaccessible to the Black Westerner. In the Indigenous borderlands, "with his dark skin a commodity whose value was heavily situational, Beckwourth was able to use, deny, and change his racial identity as he saw fit," Laura Browder argues (2000, 117). This weighed heavily in the mid-nineteenth century, as white Americans viewed Black and Indigenous people as inferior in civilization to whites, if not as distinct species altogether. Mixed-race people like Beckwourth further aroused anxieties about atavism, as whites feared the loss of "superior" white racial traits through miscegenation (Browder 2000, 22; Lape 2000, 25–26). Accordingly, Beckwourth's passing for white in his narrative represented a "hidden polemic" to those readers who knew him to be Black or mixed-race and to those whom he hoaxed into believing in his whiteness (Lape 2000, 26).

In contrast, Beckwourth's cultural appropriation of a Crow Indigenous identity in his *Life and Adventures* did not arouse public scrutiny during his lifetime. Since the twentieth century, American social discourse has come to view cultural appropriation as an illegitimate act of either claiming belonging to a marginalized group to speak on behalf of it, thus undermining notions of "authenticity," or of exploiting a marginalized culture for the benefit of a hegemonic one (Birkle 2010, 148–49; Lape 2000). However, Beckwourth's appropriation of Crow identity not only precedes this critical attitude but also follows his formal adoption into a Crow band, among whom he subsequently lived for thirteen years. To the white target readers of his autobiography, he also presents his Crow tribal membership as a cultural identity rather than an ethnic one and thus does not challenge U.S. society's ethnoracial hierarchies. This aligns with an Indigenous understanding of identity shared by the Crow that based tribal membership on cultural integration and kinship ties. Beckwourth's memoir underlines his claim to a Crow identity by depicting how members of different Native nations recognized him as both white and Crow (Beckwourth 1981, 178, 227, 378). Moreover, as Indigenous resistance against the United States' territorial expansion and rule in the West aroused white societal concern, Beckwourth's appropriating a Crow identity to contribute to the Anglo-Americanization of the West appeared justified in the eyes of his white audiences.

Although Beckwourth's racial passing and cultural appropriation elicited opposite responses among his white contemporaries, I view them as two related forms of resistance against the socially constructed boundaries restricting Black socio-spatial mobility. Beckwourth's utterance, "I learned this one truth . . . that a white man can easily become an Indian, but that an Indian could never become a white man" (323), demonstrates his awareness of, and certain adherence to, the prevailing racialized cultural hierarchy. This permits a white person to appropriate other cultures while denying others to assume whiteness. Rather than concluding that he, as a Black or mixed-race man, cannot "become Indian" or white, as Blacks were considered even less "civilized" than either other race (Haas 2022, 254), Beckwourth adopts a white and Indigenous identity in *His Life and Adventures* to obtain social and spatial mobility in the West. Although he never explicitly claims to be white in his narrative, Beckwourth never identifies as Black there, either. For instance, the narrator remarks that in his childhood, "my father removed to Missouri, taking with him all his family and twenty-two negroes" (Beckwourth 1981, 14). Contrasting the family, including "my father," and their Black slaves, the sentence distances the narrator from the latter group, Beckwourth similarly asserts his whiteness by using racist language. For example, he reproaches a young mulatto caught stealing by calling him "a black velvet-headed scoundrel" and "consummate villain." In addition to the theft, he accuses the boy of inspiring the Crow, "who are bad enough already," to commit crimes (249). This argumentation reiterates the white belief that Black or mixed-race people were even more debased than the Indigenous nations and thereby asserts the speaker's moral "superiority" based on his implicit whiteness.

Once Beckwourth has implicitly established his whiteness and assured his readers of his ongoing loyalty to Anglo-American society in his narrative, he can not only adopt a Crow identity but also strategically deploy it. Several instances in his *Life and Adventures* tellingly illustrate this: When trading at Fort Clark on the Missouri River, the narrator at first passes for a Crow: "I [was] speaking nothing but Crow language, dressed like a Crow, my hair [was] as long as a Crow's, and myself as black as a crow. No one . . . doubted my being a Crow" (Beckwourth 1981, 177). He then uses his knowledge of English and reveals his identity as a non-Indigenous trapper to obtain more goods in exchange for his wares than he would have received as a Crow. Years later, Beckwourth recounts how he drew on his renown as a Crow warrior to trade favorably with the Cheyenne. Despite a longstanding enmity among the two nations, the Cheyenne respected Crow bravery and trusted a Crow rather than any white trader (428–29, 438, 463). "I had lived so much among [Indigenous nations] that I could enter into their feelings, and be in every respect one of themselves" (463), Beckwourth explains his success. In both instances recounted here, he uses his ability to move among diverse cultures for social and economic gain and to demonstrate his intellectual superiority over his white and Native trading partners in his book. His passing for white and appropriating a Crow identity undermine the concept of race as an "inherent and unalterable essence" (Ginsberg 1996, 4) and expose its social construction and performativity, depending on geographic and socio-cultural environments and practices.

Claiming social identity through uses of nature

Beckwourth's ambivalent social identifications also show in his relation to Western nature. In his autobiography, he acknowledges the Great Plains and the Rocky Mountains as the material and spiritual sources of Indigenous livelihood. The narrator of his *Life and Adventures* repeatedly mentions the abundance of beaver and buffalo on Crow territory as well as the nation's planting pumpkin, corn, and tobacco besides trapping and hunting to provide for their subsistence, cultural practices, and trade (Beckwourth 1981, 205–07, 268, 300, 303, 343). Accordingly, he scrutinizes the decimation of the buffalo and its impact on the Indigenous nations of the West. Once numbering "by the thousand," buffalo had become "scarce" by the time of Beckwourth's life in the West (346). "The incessant demand for robes has slain thousands of those noble beasts of the prairie," the narrator criticizes and points to the futile efforts of the Crow to preserve the buffalo stock (347). Although he acknowledges the Indigenous view of the buffalo as a divine gift to sustain their peoples (529), he predicts that their reliance on this singular natural resource will be the downfall of the Native nations:

> Within half a century the race of buffalo will be extinguished from the continent. Then farewell to the Red Man! for he must also become extinct, unless he applies himself to the cultivation of the soil . . . . [The Indigenous nations] consider . . . the buffalo [to have been] sent to them by the Great Spirit for their subsistence; and when he fails them, what shall they resort to? (346–47)

Moreover, while Beckwourth's reference to the demand for buffalo robes implies the impact of the growing white trade on the nature and populations of the West, his narrator becomes more explicit in pointing to the region's physical and social transformation through growing white settlement. By the time of the buffalo's extinction, he argues,

> Much of the land [the Indigenous nations] now roam over will be under the white man's cultivation, which will extend inland from both oceans. . . . Unquestionably [the Native American] will be taken in a surround, as he now surrounds the buffalo; and as he cannot assimilate with civilization, the Red Man's doom is apparent. (347)

The passage stands out for its analogy between the Indigenous population and the buffalo as both being fatally encircled. This parallel once again reveals how the narrator shares white society's understanding of Native nations as an integral part of "savage" nature, destined for extinction if it cannot accommodate its transformation.

Beckwourth's portrayal of the landscape and the Indigenous nations of the West aligns with his professional experiences in the region (Madera 2015, 89, 102, 108) and his narrative self-fashioning as an American "frontiersman." As Judith Madera points out, his "*Life and Adventures* depicts a manifest landscape for a nationalized subject." Here, the Western "landscape is channeled into territory and territory into an exploitable form to accommodate self-interest" (2015, 88–89). Beckwourth's narrator advertises the Crow territory to his readers as suitable and desirable for white settlement. Reminiscent of the popular guidebooks for white settlers and immigrants coming to the West (Haas 2021, 52–54), he praises both the agricultural and industrial value of the area:

> Portions of it form the choicest land in the world, capable of producing any thing that will grow in the Western and Middle States. Innumerable streams . . . irrigate the plains, and would afford power for any amount of machinery . . . . The country also produces . . . wild fruit of every variety . . . and in great abundance . . . . I am fully convinced that their territory contains vast mineral wealth. (Beckwourth 1981, 347–48)

In line with the widespread Anglo-American view of nomadism as a sign of "savagery" and of Indigenous people as incapable of "properly" cultivating land, Beckwourth denies the Crow and other Native nations their right to their own territory, which he earlier defended: The Indigenous man, the narrator of the *Life and Adventures* argues, "is a vagrant, and he must wander; he has no associations to attach him to one spot; he has no engendered habits of thrift or productiveness to give him a constant aim or concentration of purpose" (324). In contrast, Beckwourth foregrounds his and other mountain men's labor to prepare the West for white appropriation. His narrator states, "the forts that now afford protection to the traveler were built by ourselves at the constant peril of our lives . . .; without well-furnished wagons to resort to when hungry; no roads before us but trails temporarily made" (51–52). In so doing, he presents the land as a resource and its transformation into a usable commodity as a non-Indigenous vanguard's heroic endeavor. His narrative thereby "undermines Indigenous possession and cedes to dominant territorial codes of representation" (Madera 2015, 89).

Despite the narrator's admitted role in bringing forth the transformation of Western nature and the demise of the region's Indigenous civilizations, Beckwourth's "*Life and Adventures* also relays Indian knowledge to subverting white normative territorialization," Madera observes (2015, 104). On the Great Plains and in the Rocky Mountains, both natural and human-made landmarks such as rivers, canyons, Anglo-American forts, and Indigenous villages provide spatial orientation in this vast, sparsely settled region. Although the text commonly uses place names coined by white explorers, it occasionally adds the respective Native names. Even more, it primarily identifies large stretches of land by the Indigenous populations residing there. As all Plains nations were at least semi-nomadic, they did not draw borders across their lands (Carlson 1998, 111). In legitimizing mobile terrains and alternative names, Beckwourth's narrative undermines the Anglo-American practice of preparing territorial expansion through naming and mapping the terrain. The text thereby renders readers aware of the powerful and diverse Indigenous presence in the region that counters nineteenth-century popular images of an empty "virgin land" waiting to be taken (Haas 2021, 53) as much as subsequent perceptions of the region based on its American political division and white settlement structures.

Conclusion

As the present essay shows, studying the life and autobiography of James Beckwourth provides valuable insights into Black narrative self-making in the North American West during the first half of the nineteenth century. This complements existing scholarship in autobiography research and Black Indigenous Studies, which hitherto have largely ignored Beckwourth's narrative. In his life and autobiography, Beckwourth subverted racial, spatial, and narrative norms and regulations to achieve social agency and economic opportunities commonly denied a Black man in the mid-nineteenth-century United States. In his memoir, he justifies his portrayal of the West as a violent place and his own contribution to this violence with the educational purpose of informing adventure-seeking readers about the hardships and perils of life in the region (Beckwourth 1981, 534). By setting his memoir in the formally and thematically heterogenous mold of Western autobiography in a manner that also positively asserted racial passing and cultural transformation, he enabled his complex and contradictory self-fashioning as an implicitly white American expansionist Crow warrior that undermines essentialist and hierarchical notions of racial, cultural, or national identities.

Beckwourth's political emancipatory agenda was limited, though, by how his actions and writing "simultaneously threaten[ed] Crow survival and praise Crow humanity and ingenuity" (Lape 2000, 21). Although he validates Indigenous lifeways in his autobiography, he also exploits them for his personal and white America's economic and political gains. Through his life in the West, Beckwourth established himself as an (implicitly white) American "frontiersman" who showed his loyalty to his country by strategically appropriating an Indigenous identity. Given this record of using his life among the Crow as a resource to claim whiteness and Americanness, it is highly ironic, though, that his contribution to advancing Anglo-American civilization in the West threatened not only the Plains Indigenous nations but also the lifeways of men like himself.

Works cited

Adelman, Jeremy, and Stephen Aron. 1999. "From Borderlands to Borders: Empires, Nation-States, and the Peoples in Between in North American History." *American Historical Review* 104, no. 3: 814–41.

Beckwourth, James P. 1981. *The Life and Adventures of James P. Beckwourth, as Told to Thomas D. Bonner*. 1856. Ed. Delmont R. Oswald. Lincoln, NE: Bison Books. Orig. pub. 1856.

Birkle, Carmen. 2010. "Mediation and Appropriation: Imaginary Indians and the Grey Owl Syndrome." In *Auto/Biography and Mediation*, ed. Alfred Hornung, 139–59. Heidelberg: Winter.

Browder, Laura. 2000. *Slippery Characters: Ethnic Impersonators and American Identities*. Chapel Hill: University of North Carolina Press.

Carlson, Paul H. 1998. *The Plains Indians*. College Station: Texas A&M University Press.

Couser, G. Thomas. 2004. *Vulnerable Subjects: Ethics and Life Writing*. Ithaca, NY: Cornell University Press.

Doolen, Andy. 2014. *Territories of Empire: U.S. Writing from the Louisiana Purchase to Mexican Independence*. New York: Oxford University Press.

Gerhardt, Christine. 2002. "The Greening of African American Landscapes: Where Ecocriticism Meets Postcolonial Theory." *Mississippi Quarterly* 55, no. 4: 515–33.

Ginsberg, Elaine K., ed. 1996. "Introduction: The Politics of Passing." In *Passing and the Fictions of Identity*, 1–18. Durham, NC: Duke University Press.

Haas, Astrid. 2021. *Lone Star Vistas: Travel Writing on Texas, 1821–1861*. Austin: University of Texas Press.

Haas, Astrid. 2022. "Native Bondage, Narrative Mobility: African American Accounts of Indigenous Captivity." *Journal of American Studies* 56, no. 2: 242–66.

Hämäläinen, Pekka. 2008. *The Comanche Empire*. New Haven, CT: Yale University Press; Dallas: William P. Clements Center for Southwest Studies, Southern Methodist University.

Hoxie, Frederick E. 1995. *Parading through History: The Making of the Crow Nation, 1805–1935*. New York: Cambridge University Press.

Isenberg, Andrew C. 2004. "Environment and the Nineteenth-Century West: Or, Process Encounters Place." In *A Companion to the American West*, ed. William Deverell, 77–92. Malden, MA: Blackwell.

Krupat, Arnold. 1981a. “American Autobiography: The Western Tradition.” *Georgia Review* 35, no. 2: 307–17.
Krupat, Arnold. 1981b. “The Indian Autobiography: Origins, Type, and Function.” *American Literature* 53, no. 1: 22–42.
Lape, Noreen Grover. 2000. *West of the Border: The Multicultural Literature of the Western American Frontiers*. Athens, OH: Ohio University Press.
LeMenager, Stephanie. 2003. “Trading Stories: Washington Irving and the Global West.” *American Literary History* 15, no. 4: 683–708.
Madera, Judith. 2015. *Black Atlas: Geography and Flows in Nineteenth-Century African American Literature*. Durham, NC: Duke University Press.
Mays, Kyle T. 2021. *An Afro-Indigenous History of the United States*. Boston: Beacon Press.
Nugent, Walter. 2008. *Habits of Empire: A History of American Expansion*. New York: Knopf.
Oswald, Delmont R. 1981. “Introduction” and “Epilogue.” In *The Life and Adventures of James P. Beckwourth, as Told to Thomas D. Bonner*, ed. Delmont R. Oswald, vii-xiii, 603–28. Lincoln, NE: Bison Books.
Ravage, John W. 1997. *Black Pioneers: Images of the Black Experience on the North American Frontier*. Salt Lake City: University of Utah Press.
Roberts, Alaina. 2021. *I’ve Been Here All the While: Black Freedom on Native Land*. Philadelphia: University of Pennsylvania Press.
Sayre, Gordon M. 2000. “Introduction.” In *American Captivity Narratives: Selected Narratives with Introduction*, ed. Gordon M. Sayre, 1–17. Boston: Houghton Mifflin.
Smith, Sidonie, and Julia Watson. 2010. *Reading Autobiography: A Guide for Interpreting Life Narratives*. 2nd ed. Minneapolis: University of Minnesota Press.
Taylor, Quintard. 1998. *In Search of the Racial Frontier: African Americans in the American West, 1528–1990*. New York: Norton.
Wilson, Elinor. 1988. *Jim Beckwourth: Black Mountain Man and War Chief of the Crows*. 4th printing. Norman: University of Oklahoma Press.
Zafar, Rafia. 1997. *We Wear the Mask: African Americans Write American Literature, 1760–1870*. New York: Columbia University Press.

Ines E. Veauthier

Bridging Worlds: Individual and Collective Reflections

This paper takes a closer look at what might be called a glimpse into the lives of two Nobel Prize laureates, namely the Canadian writer Alice Munro and the U.S.-American author John Steinbeck. These glimpses into individual lives will be examined using the historical and cultural background provided by the writers in order to demonstrate their ways of communicating social issues. Their strategies and techniques for connecting and relating their memories will also be discussed.

The idea of 'bridging worlds' points at some of the challenges of life writing and also summarizes the theses of the paper: thoughts, actions, or experiences can only be understood if you provide the respective context because the historical or cultural circumstances will have an impact on developments. Generally speaking, the major codes of society are culturally embedded and taken for granted (Nichols 1981, 30). The members of a community do not usually become aware of the role these codes play in creating reality. By learning and adopting the values of one's family, friends, and cultural environment, the individual internalizes the codes the society endorses. This means that the concept of individual identity is seen within the framework of collective memory and social change.

The analysis aims to prove that Munro and Steinbeck act as mediators and guideposts. Their experiences "need to be translated across multiple cultural gaps" (Costantino 16). Talking about their lives also takes readers along on a journey to understand that one's social reality does not have the status of a natural truth but is subject to change over time and space. Likewise, a 'search for America' will not only provide the outsider's view but will also entail being part of the community. Prior to this analysis, a closer look at the authors discusses their connections. A brief summary of some relevant aspects of life writing establishes the context required for an in-depth examination of the primary sources. Keeping in mind that the authors want to connect people, it will be assessed if their writing "rings true" (Adams 16), i.e. feels authentic instead of looking for perfect accuracy in the autobiographical accounts.

Despite the fact that Steinbeck and Munro are both Nobel Prize laureates, comparing them might not be an obvious choice because at first glance they seem too different: Steinbeck was born in 1902 in the U.S.A., and when Alice Ann Laidlaw was born in Canada almost a generation later, in 1931, Steinbeck had already published a novel and several short stories (the following sources were used to compile this synopsis: Howells 2017; McGrath 2012; McWilliams 2007; Österling

https://doi.org/10.1515/9783111551968-007

1969; Schultz and Li 2005; Serafin 2022; Steinbeck 1962; The Steinbeck Institute; Thacker 2022). While Steinbeck's career started in his late twenties, Alice Munro published her first story when she was nineteen and her first collection of short stories when she was in her late thirties, in 1968, the year of Steinbeck's death.

However, a closer look reveals that the authors have a lot of things in common. Some of these aspects may seem disconnected but putting them all into context will reveal that these are defining factors for Steinbeck and Munro alike, proving that they are connected in their words and actions.

To begin with, the texts to be examined were written after a long writing career and shortly before receiving the Nobel Prize. In his acceptance speech, Steinbeck referred to William Faulkner; Munro's work and her locations in Canada are often compared to Faulkner's writing. The authors were honored with numerous prizes, and some of their writings were adapted into films, yet, neither Steinbeck nor Munro enjoyed the celebrity status much.

Both of them grew up in a rural environment and a distinct sense of place is part of their writings. Their childhood and the chores they had also shaped their general idea of animals having a certain function for humans. Both, Olive Hamilton Steinbeck and Anne Clarke Chamney Laidlaw, were former school teachers. When the mother was in bad health, John or Alice, respectively, would take care of her. John and Alice grew up in a nuclear family with several siblings, and some of their ancestors were from Ireland. Their teenage years were overshadowed by global conflicts as World War I began when Steinbeck turned 12 and WW II ended when Munro was 14. Steinbeck enrolled at Stanford University, listing English as his major field of study, but he never graduated, similar to Munro who studied English and journalism at the University of Ontario without graduating (but she received an honorary degree in 1976). Both were divorced and remarried. These biographical parallels translate into the background for their ambitions, social criticism, and writing style.

While Steinbeck could hardly have noticed the Canadian writer who was much younger and relatively unknown when he was alive, Munro was not only widely read but also opened a bookstore with her husband in 1963. It cannot be established whether John Steinbeck's works had any influence on her oeuvre. However, critics have compared her writing to a number of well-known authors, such as Charlotte Brontë (Garson 2000), Thomas Hardy (Beran 1999), Gustave Flaubert (Giordani 1995), Anton Chekhov (McGrath 2012), Willa Cather (Stich 1989; Thacker 1992), James Joyce (Martin 1979; McWilliams 2007), William Faulkner (Gault 2006), Margaret Atwood (Marra 1996; Wall 1997), Toni Morrison (Frye 1986), – and also John Steinbeck (Carrington 1994). The analysis will prove that Steinbeck's and Munro's autobiographical writings have connections that run deeper than the surface and that they both act as mediators, illustrating "the role of memory or recollection as a medium of time and reality" (Hornung 2010, xii). In *Travels with Charley*, Steinbeck speaks to the audience in an innovative, cultural dialogue; Munro reflects on her childhood and uses these memories for both historical and cultural mediation.

The subsequent outline regarding life writing will provide the necessary context. Telling others all about your life sounds like a challenge that can never be successfully met as there are too many obstacles in the way. Two of the most obvious issues would be what you will talk about and how you can communicate it. As it seems unrealistic to relate every single moment of one's life, it will be necessary to decide what is worth mentioning. Being selective about the topics is mirrored by the choice of styles and strategies used to narrate the memories. These factors may explain why auto/biographical writing remains complex and multifaceted despite the long tradition both of the genre and of its reception (Kendall 2022). The choice also depends on the receiving party and their interpretation of the life writing and what they can relate to. This illustrates the point that writing is part of the ongoing process of bonding because "all literary works . . . are 'rewritten,' if only unconsciously, by the societies which read them" (Eagleton 1983, 12).

The relevant terms themselves are not unified, so there are various answers to the question of which criteria need to be met by a text to qualify as auto/biography, life narrative or memoir, to name just a few (Howes 2020). For the purpose of this analysis, the concept of life writing as an umbrella term and as a concept of "openness and inclusiveness across genre" promises the best results (Jolly 2001, ix) while using "auto/biography" as a quasi-synonym. Deciding what counts as life writing will be discussed again when examining John Steinbeck's *Travels with Charley*. Alice Munro explicitly defines her vignettes "The Eye," "Night," "Voices," and "Dear Life" as a glimpse into her life while stating they might be "autobiographical in feeling, though not, sometimes, entirely so in fact" (Munro 2013, 255).

On the one hand, this emphasizes that individual texts often defy clear labeling. They might belong to different subcategories such as thematic, religious, intellectual types or special-purpose auto/biographical writing (Kendall 2022). Indeed, a more radical view describes elements reflecting one's life as the basis permeating any and all thinking and writing. Seen like that, auto/biography can be defined not so much as a genre but as "a figure of reading or of understanding that occurs, to some degree, in all texts" (De Man 1979, 921). This argument can even be interpreted as intertextuality providing an element of connection among texts and thus also among the authors who wrote them. Depending on the target group and the intentions of the authors, it may be possible to see life writing as a way to connect people.

On the other hand, there is a conflict between looking back at one's life and finding a way to present a true account based on facts. Outsiders might accuse the writer of creating a fairy tale by leaving out unflattering details and painting a picture-perfect world. Munro's comment shows that she seems to have anticipated this kind of criticism. The study of auto/biography confirms several reasons why there is a close relationship with fiction – epitomized in the term autofiction (Dix 2022).

Life writing means looking back at the decisions, circumstances, and dreams that constitute one's life. Everything can be reassessed, thus finding some deeper meaning in what happened. On the one hand, reflected memory will always entail gaps, imprecision, and deformation of what is perceived as facts because life needs to be told in retrospect. On the other hand, the medium itself contains errors if you understand life writing as "the mirror image of a life" (Gusdorf 1980, 40) or as a "self-portrait" (Howarth 1980, 85). Moreover, the individual experiences may also reflect the historical time and society to which the writer belongs, so the framework of collective memory plays a role. Three well-established modes of approaching "self-portraits" need to be mentioned for the subsequent analysis: first, autobiography as oratory when the authors follow religious or political aims "to carve public monuments out of their private lives" (92). Second, if a writer prefers entertaining the audience, life is depicted as drama or as "a staged performance" (96). Third, autobiography as poetry is characterized by an experimental, self-critical style and the authors seem to reserve "the right to change their ideas" (106). The aesthetic qualities of language seem more important than convincing others of higher aims or staging events. Also, the text may belong to more than one category as the writer's focus changes, i.e. "the narrating 'I' may be multiple, different 'I's" (Smith 2019, 77).

While facts and fiction may be hard to distinguish, one suggestion is that autobiographical truth means the writer is authentic, and the story provides meaning (Adams 1990, 167–69). Thus, any well-written text can blur the boundaries of being honest or stretching the truth: "There is no specifically linguistic essence of fictionality that is immediately perceptible in the particulars of a text" (Foley 1986, 40).

Munro's first important memory seems to be her emotional coming-of-age as a five-year-old in the vignette "The Eye," similar to the typical concept of a childhood auto/biography that starts "with the first light of consciousness" (Coe 1984, 9). Munro notices that the sensations declared by her mother do not feel true: "Up until the time of the first baby I had not been aware of ever feeling different from the way my mother said I felt" (Munro 2013, 257). The girl starts adjusting her emotions according to what pleases her mother (258) but she is becoming more detached and independent, confiding "I began to accept how largely my mother's notions about me might differ from my own" (258). This corresponds with the current research on children's development: "Between 18 and 30 months, individuation (autonomy) emerges. . . . Between 30 and 54 months . . . [t]esting limits on what behaviors are acceptable and how much autonomy they can exert is an expected phenomenon" (Malik and Marwaha 2022, n. pag.). Accordingly, the child's surprise at having her own feelings is complemented by the realization that it had better remain a secret: "I knew enough not to speak about this to anybody" (Munro 2013, 258).

This is also an instance of Munro's confessional strategy when she admits a certain subversive streak: "And something in me was turning traitorous, though she didn't know why, and I didn't know why either" (Munro 2013, 263). Studies on the experience of childhood confirm that mentioning one's rebellious past is a typical approach: "The confession implies at least a strong element of continuity (a continuity of responsibility, if of nothing else) between child and adult" (Coe 1984, 75).

Munro's main strategy is closely connected. She manages to interact with her audience by using metalepsis which is traditionally employed by authors to merge worlds, "a crossing of the boundary between narrative and reality" (Budelmann 2020, 59). Readers can thus feel less like observers and instead get closer to the fictional characters or can 'live' in the past. Trying to pull people into her narration, Munro uses a guiding voice that develops from a side comment to herself – "Why do I say ominous?" (2013, 257) – to a direct turn towards her audience explaining her past, "Jesus suffering the little children to come unto him. Suffering meant something different in those days" (258). The adult wants to confirm that she is not making things up while conceding that her memory might be incomplete. She resorts to authentication methods most people use in daily life, relating one incident to another (Bergmann 1998, 111), thus creating an illusion of objectivity: "And were the refreshments really as lavish as I remember? With everybody so poor? But maybe they were already feeling not so poor, with the war jobs and money that soldiers sent home. If I was really ten, and I think I was, then those changes would have been going on for two years" (Munro 2013, 291).

Munro stresses the truthfulness of her memories when she explains the relationship between facts and fiction, describing a woman who turned out to be a prostitute: "I think that if I was writing fiction instead of remembering something that happened, I would never have given her that dress. A kind of advertisement she didn't need" (Munro 2013, 292). She feels the need to explain that she did not make anything up, but "this is not a story, only life" (307). Her readers are told that because of the contradictions it is the truth: "It wouldn't do in fiction" (309).

Some of these explanations may fulfill more than one function; while they can be interpreted as social criticism, the author also uses this technique to act as a cultural mediator. To name just a few examples, she mentions the general hygienic standards (or sub-standard) of the public toilets at school (Munro 2013, 300–301). There is also a discussion about the social status and the aspirations expected of members of society in the 1940s: only about 25% of the students would stay in high school until grade thirteen. Most of them dropped out before, getting jobs or having babies (303), compared with more current statistics showing that 52.1% of young men and 70.2% of young women have a college certificate or a diploma or a Bachelor's degree (Statistics Canada, 2019). Munro also describes the distinctions due to one's social class, particularly the idea of farmers vs. townspeople. Her mother, for example, was judged badly because despite having been born and raised on a farm she became a teacher and longed to find more in life than what was dealt out to her at birth, but she "spoke in such a way that her own relatives were not easy around her" (Munro 2013, 304).

Several taboos are explained, some of which she only understood step by step herself. Nobody openly talked of cancer: "The only way I can explain our failure to speak of it was that there must have been a cloud around that word like the cloud around the mention of sex" (Munro 2013, 273). Despite the taboos, she experienced sexual harassment in a very explicit way: "It appeared that the same boys who'd always asked me rhetorically and alarmingly if I wanted to fuck were just as eager to get jobs as their older brothers were to go into the Army" (300). The boys' outspokenness is particularly incongruent when comparing these verbal attacks with the complete avoidance of the topics of sex and prostitutes by her family and others around her.

In Munro's account, the outcasts of society are female. As a young girl, she was not allowed to see her friend anymore and later found out the reason: "Diane's mother had been a prostitute and had died of some ailment it seemed that prostitutes caught" (Munro 2013, 302). All the questions she must have had regarding this ailment in particular and prostitutes or sex in general are left unanswered because nobody talks about these topics and there is no one she can ask. At this point Munro leaves her readers in the dark so they can feel the way she felt in that situation.

Likewise, in the episode "Voices" the girl understands only in hindsight that her mother wanted to leave the country dance as soon as she spotted a certain woman. She was a "disgrace" and because of her, "it was all ruined" (295). If the incident had to be mentioned at all, only euphemisms would be used: "And I would not have used the word prostitute. Bad woman, more likely" (292).

One woman at Sadie's wake remarked: "A girl without a boyfriend going to dances on foot" was "asking for trouble" (Munro 2013, 267). This real-life incident seems to echo the literary tradition of women being confined to strict gender roles or else finding an escape only through death or insanity – until Dreiser's *Sister Carrie* (1900[1998]) established a new vision for women's emancipation when the heroine "goes unpunished for her transgressions against conventional morality" (Kuiper 2022, n. pag.).

The treatment of children is another aspect needing a historic or cultural explanation. From today's point of view, raising a child with regular beatings is cruel and criminal, and people will lose custody of their child if social services hear about such practices. Munro emphasizes that the physical punishment or "strappings" were deemed "no more than the necessary and adequate curbing of a mouthy child's imagining that she could rule the roost" (2013, 284). Her final evaluation of another example of childcare shows how individuals are influenced by the society they live in. She feels certain she would find an expert to help a child suffering from problems of sleeplessness and delusions, given that at the time of writing she lives in a society upholding different values of education, socialization, and a child's psyche. But she also concludes that her father managed to help her when she had severe problems: "However, on that breaking morning he gave me just what I needed to hear" (284).

Munro takes her readers back in time, explaining the circumstances of her childhood and family life that might feel unfamiliar. This can be interpreted as a general strategy to understand yourself and the world around you in a cross-cultural approach (Decker 2016, 215). If 'culture' is meant to comprise changes over time, it can also serve to explain Munro's attempt to add a historic perspective by providing the social context of her memories. It may be concluded that Munro wanted to include this part of her childhood. However, anyone who did not grow up in the 1930s and 1940s would miss out on the general context needed to fully understand the situations. From today's point of view, readers might interpret Munro's past as disturbing, violent, and characterized by child abuse. Her explanations serve to provide the context and thus greater understanding according to the theory that "insight, meaningfulness, and contemplations of the human condition may be important gratifications offered by entertainment" (Oliver and Woolley 2011, 138). Analyzing her use of metalepsis with its enhanced reader involvement, it can be concluded that her narration techniques contribute to a form of social action and change through language. Her writing thus exemplifies the performativity of narration (Häsner 2001, 20).

Munro sees herself as the protagonist of Lucy Maud Montgomery's 1908 novel and when she remarks that she tries to imitate the protagonist's lifestyle she is implicitly referring to Oscar Wilde's discussion of the relationship between art and real life: "But I was used to this and could easily ignore it all, constructing for myself something out of the books I liked, such as *Anne of Green Gables* or *Pat of Silver Bush*. . . . Fresh manure was always around, but I ignored it, as Anne must have done at Green Gables" (Munro 2013, 305). On the one hand, this cultural allusion positions Munro within the literary tradition of Canadian women authors (Howells 2017, 217–38). On the other hand, this technique of intertextuality blurs the boundaries between reality and fiction similar to the way in which her metalepsis transgresses boundaries between narrative levels.

The vignette "Voices" demonstrates a different approach. Analyzing how the events are dramatized in retrospect in Munro's imagination, it is as if she puts the action on the stage for the audience. Therefore, it resembles the contemporary idea of a soap opera which again relies on well-established formats as "melodrama builds on underlying oral structures, formulas, and archetypes that [have been] shared by cultures" and this has developed over time and through verbal, print, and broadcast forms (Straubhaar and La Pastina 2005, 275). While the grown-ups understand that Peggy is a prostitute, the ten-year-old Alice feels awed and identifies with the heroine in the drama unfolding before her: "Nevertheless the young men treated [Peggy] as if she was someone who deserved never to have encountered one rough moment, someone who rightfully should be petted and pleasured and have heads bowed before her" (Munro 2013, 295). The effect of this kind of storytelling is an emotional connection as the audience can relate to the events. This can be termed the "multi-layered 'cultural proximity' of the melodrama genre" (Ríos and Castañeda 2011, 4). It is also a method of mediating between self and community on the one hand and reality and fiction on the other.

Munro explicitly calls her memories autobiographical vignettes. She is equally open about the interaction with her readers which is reflected in her writing style. The subsequent analysis will show how a different writer who is more vague about his intentions complements Munro's writing. Their stories can be seen as an intertextual conversation or a duet of two unique voices trying to make people listen.

John Steinbeck never claimed *Travels with Charley* to be an autobiography. Nevertheless, his travelogue contains his memories and thoughts regarding a certain time of his life and besides, "a man's lifework is his fullest autobiography and, he being what he is and where and when he is, neither the lifework nor the autobiography could be otherwise" (Olney 2017, 3).

The book was criticized for deviating from a strictly factual, statistical account: "The fictional aspects of *Travels with Charley* are noticeable on most pages, the chief of these being the use of dialogue" (Parini 2000, xv). Steinbeck claims it is a "little book of ambulatory memoirs" (1975, 702) and some critics interpret it as a mythic journey, a "novel-like odyssey of a beloved American writer and his dog" (Hughes 2013, 316). Steinbeck himself understands the ambivalent relationship between facts and fiction yet he aims for meaningfulness and authenticity as discussed above, the "small diagnostic truths which are the foundations of the larger truth" (Steinbeck 2000, 5). Moreover, he addresses potential criticism and the risk of overinterpreting statements; he remembers some notes stuck on a bottle he wrote for later inspection that read "Relationship Time to Aloneness," and "Reversion to pleasure-pain basis," explaining which thoughts are summarized in these cryptic notes. Later on, he adds a subversive comment: "Good Lord, the trails we leave! Suppose someone, finding the ketchup bottle, tried to reconstruct me from my notes" (106–07).

At first glance, *Travels with Charley* can be categorized as an epic journey, a monomyth or a heroic quest featuring departure, initiation, and return both in a literal and a more abstract way as "beyond the protection of his society [there is] danger to the member of the tribe" (Campbell 1973, 64). Steinbeck, however, adds a twist confiding that the journey has a personality of its own (Steinbeck 2000, 3).

Three main aspects will be analyzed. The first one concerns Charley as a non-human companion, serving as a sounding board for Steinbeck's impressions. This can be interpreted as a technique to engage the audience (Hughes 2013, 314). While Steinbeck encounters many people, they all have their own viewpoints, so there is no access to an objective, all-encompassing 'American' truth. However, Charley embodies an entity of his own that is not subject to the conventions of society. There is an echo of cultural competencies when Steinbeck emphasizes that his companion is bilingual – even if English means "he has to translate, and that slows him down" (Steinbeck 2000, 7). A dog in general acts as "a bond between strangers" (8), and Charley's status as a mediator is emphasized by his skills as a communicator and "mind-reader" who "has always associated with the learned, the gentle, the literate, and the reasonable both in France and in America" (106).

Charley is also characterized as a "tree expert of enormous background" (Steinbeck 2000, 143). Steinbeck creates an episode involving sequoias and their sacrosanct status and he is strongly disappointed when Charley does not seem to share his reverence for the trees (144–45). In this situation the quest merges with Charley's function: "This was the time I had waited for. I opened the back door and let Charley out and stood silently watching, for this could be dog's dream of heaven in the highest. . . . 'Look, Charley. It's the tree of all trees. It's the end of the Quest'" (144). This may be understood as an ironic comment on the perception of a journey as the quest of a mythical hero. Steinbeck clearly subverts this view of his project.

Charley is presented as a companion who is almost better than a human friend because the author can project his own ideas onto him. In conclusion, this supports the thesis that from his childhood on, Steinbeck would expect animals to have a certain function. While this may have happened only subconsciously for Steinbeck, the field of dog-assisted therapy has meanwhile become an established option (Hüsgen, Peters-Scheffer, and Didden 2022).

The second aspect concerns ideals and values which, according to one of Steinbeck's friends, would be hard to find: "I haven't seen anything but cowardice and expediency. This used to be a nation of giants. Where have they gone?"

(Steinbeck 2000, 128). Apart from a lack of giants, the author notices a growing tendency toward materialism or consumerism: "The mountains of things we throw away are much greater than the things we use" (22). An alternative approach would be deciding what is really important in life and acting accordingly which calls to mind Thoreau's maxim: "Simplify, simplify" (Thoreau 1854, 65). Steinbeck's motivation for his journey can be compared with Thoreau's quest in the woods: "The work of Walden was to model a way of life that Thoreau believed was truer to American ideals than that surrounding him in antebellum America" (Couser 2021, 16). Meeting a young man in the countryside and discussing his dreams and ambitions (Steinbeck 2000, 131–34) shows on the one hand that the younger generation might be more willing to sacrifice safety and comfort in order to be true to their ideals. On the other hand, there is a strong urge to conform to traditions and social pressure, to submit to the collective 'wisdom' of parents and educators. This brings us back again to Thoreau and the revision of the master-slave paradigm when he claims it is "worst of all when you are the slave-driver of yourself" (Thoreau 1854, 8). In general, however, changes in society are unlikely to happen as long as there is no real pressure: "Beliefs and practices tend to persist not because they are optimally beneficial but because they generally work just well enough that changes in them are not self-evidently needed" (Edgerton 2000, 137).

An approach that might be termed the soul of things is connected with the above topics. It includes the importance of context, belonging or of identity within a community of spirituality. Steinbeck worried that people might recognize his name or face and would not be natural and authentic anymore. However, nobody realized who he was, so he remained incognito because "people identify things only in context" (Steinbeck 2000, 6). His own need to create a context when interacting with people and be a part of the whole also points to the limits of his analysis. Leaving Monterey, Steinbeck realizes he cannot write a simple, basic truth about America. Moreover, he cannot examine facts in a pure and sterile way that would be objective and impersonal because his own involvement prevents the neutral view of an outsider: "This monster of a land, this mightiest of nations, this spawn of the future, turns out to be the macrocosm of microcosm me. . . . From start to finish I found no strangers" (159). This thought of an inner connection sounds like an echo of an idea that has been influential in Anglo-American culture since John Donne phrased it in 1627: "No man is an island, entire of itself – every man is a piece of the continent, a part of the maine" (Abrams and Greenblatt 2006, 1305). Steinbeck's intertextual reference illustrates how the idea of human bonds is anchored in society.

Accordingly, the last aspect to be discussed is Steinbeck's strategy of bonding and the subversion and reconfirmation thereof. At the beginning of his journey, he explains that pretending to be lost is a good way of getting to talk with strangers (Steinbeck 2000, 8). Sometimes he truly gets lost, e.g. when he is planning to go to Sauk Centre to visit Sinclair Lewis's birthplace. This time his "techniques of opening conversation" (8) seem to backfire because the people he asks for directions ridicule him (101) and also prove to be completely ignorant of Sinclair Lewis's importance as one of the *lieux de mémoire* of their community (102).

In the very last scene, however, Steinbeck turns full circle as he cannot seem to find his way home in his own town anymore. This time his desperation is not a superficial strategy, so his authentic emotion provides a true connection with the person he encounters, a police officer: "He grinned happily. 'Think nothing of it, Mac,' he said. 'I got lost in Brooklyn only Saturday. Now where is it you were wanting to go?'" (Steinbeck 2000, 210). Structuring his memories in this particular way can be considered a framing technique. Moreover, it is also a subversive reference to the monomyth in which the hero will return home after his adventures.

Steinbeck wants to be a mediator and build bridges, and perhaps he even strives for "reconciliation" (Eakin 2020, 57). While his writing style is different from Munro's, it is possible to compare their bonding strategies. Munro uses metalepsis throughout to address her readers; Steinbeck gives the audience the chance to 'watch him think' as he unfolds his impressions and thoughts seemingly to himself or often directed at Charley, his sounding board. His soliloquies have the effect of granting a glimpse into his inner world which is similar to Munro's metalepsis.

Conclusion

Trying to put Munro's personal narrative into clear categories shows that her writing is mostly poetic or experimental if using Howarth's criteria outlined above. In the episode "Voices," a tendency for a dramatic approach is noticeable and overall a certain spiritual aim can be detected. These both serve to establish a connection with her audience. Likewise, she uses a confessional tone and metalepsis throughout her memories. She provides comments and explanations in order to ensure the readers, who are considerably younger or are unfamiliar with the cultural context, actually understand.

Steinbeck's autobiographical travelogue can be classified as predominantly poetic or experimental because it seems important to him to express his ideas well. According to Howarth, some dramatic elements can be detected when Steinbeck wants to entertain the readers, using irony or exaggeration. When talking about nature he uses a certain spiritual tone. These strategies may reflect that he changes his focus from time to time and also that he wants to provide bridges to reach out to people from all walks of life.

Both of them show that art can create everlasting impressions, the "intensifying and immortalizing of the moment" (Assmann 2012, 124). If the idea of bridges and connections with people is seen as the overall purpose of their writing, a certain ambivalence needs to be considered in this context. On the one hand, human relationships have the potential to support and enrich. As researchers at Harvard University found out, "close relationships . . . are what keep people happy throughout their lives [and] help to delay mental and physical decline" (Mineo 2017, n. pag.). On the other hand, relating to other humans involves more than the positive aspects of belonging and enjoying the recognition within a community. There are two sides to human interactions so "we are not only constituted by our relations [to others] but also dispossessed by them as well" (Butler 2004, 24). Accordingly, alterity concepts have to be taken into account, too, and that involves the chances of alienation and rejection. Steinbeck's and Munro's personal narratives successfully handle potential conflicts and contested views by creating bridges and by connecting people and cultures, proving that "auto/biographies are involved in literary, cultural, psychological, legal, or political processes of mediation" (Hornung 2010, xii).

Auto/biographical writing may be seen as a contested genre because of the different views of facts and fiction or the unrealistic demand for a complete and objective account of a life. At the same time, this points out the great potential of the genre. Munro's feelings about writing about her own life as mentioned above illustrate the aspect of providing an individual interpretation or impression: "Anchored in the truth of fact, though seeking the truth of interpretation, biography tends to be more stable than other literary arts" (Kendall 2022, n. pag.). Moreover, there is an aspect of appreciating the many facets of potential lives. If it is possible to write about one's memories without having to adhere to a standard approach and a strict procedure, then this individualized way of living becomes an option for others: "And this suggests to me that, collectively, memoir can play a democratizing role in our culture, bringing more and more lives – and kinds of lives – to light" (Couser 2021, 20). To add another dimension, these individual lives can be seen as a reflection of society as a whole so that the collective identity will be influenced by the individual images as voiced and visualized in life writing such as Munro's vignettes "The Eye," "Night," "Voices," and "Dear Life" and Steinbeck's personal narrative *Travels with Charley in Search of America*.

Understanding things in context means knowing what it takes to be a member of a particular society. Since "culture can be defined as a kind of 'macro-code,' consisting of the numerous codes which a group of individuals habitually use to interpret reality" (Danesi 1994, 18), the importance of culture and of mediators like Munro and Steinbeck is highlighted.

Works cited

Abrams, Meyer Howard, and Stephen Greenblatt, eds. 2006. *The Norton Anthology of English Literature*. New York: Norton.

Adams, Timothy Dow. 1990. *Telling Lies in Modern American Autobiography*. Chapel Hill: University of North Carolina Press.

Assmann, Aleida. 2012. *Introduction to Cultural Studies. Topics, Concepts, Issues*. Berlin: Erich Schmidt.

Beran, Carol. 1999. "Thomas Hardy, Alice Munro and the Question of Influence." *American Review of Canadian Literature* 29, no. 2: 237–58.

Bergmann, Jörg. 1998. "Authentisierung und Fiktionalisierung in Alltagsgesprächen." In *Inszenierungsgesellschaft. Ein einführendes Handbuch*, ed. Herbert Willems and Martin Jurga, 107–23. Opladen: Westdeutscher Verlag.

Budelmann, Felix, 2020. "Metalepsis and Readerly Investment in Fictional Characters: Reflections on Apostrophic Reading." In *Metalepsis: Ancient Texts, New Perspectives*, ed. Sebastian Matzner and Gail Trimble, 59–78. Oxford: Oxford University Press.

Butler, Judith. 2004. *Precarious Life. The Powers of Mourning and Violence*. London: Verso.

Campbell, Joseph. 1973. *The Hero with a thousand Faces*. Princeton: Princeton University Press.

Carrington, Ildikó de Papp. 1994. "Talking Dirty: Alice Munro's 'Open Secrets' and John Steinbeck's *Of Mice and Men*." *Studies in Short Fiction* 31, no. 4: 595–606.

Coe, Richard Nelson. 1984. *When the Grass Was Taller – Autobiography and the Experience of Childhood*. New Haven: Yale University Press.

Costantino, Manuela, 2010. "The Sensational Life Story of Latifa: Mediating 'Forbidden' Knowledge and Identities." In *Auto/Biography and Mediation*, ed. Alfred Hornung, 15–20. Heidelberg: Winter.

Couser, Thomas. 2021. *The Work of Life Writing: Essays and Lectures*. New York: Routledge.

Danesi, Marcel. 1994. *Messages and Meanings: An Introduction to Semiotics*. Toronto: Canadian Scholars.

Decker, Jan-Oliver. 2016. "Sich über sich selbst verständigen – sich selbst verstehen. Selbstreferenz und Selbstreflexion in der Literatur aus narratologischer Perspektive: Eine Einführung in den Band und ein Vorschlag zur Analyse und Funktion der Metalepse." *KODIKAS/CODE. An International Journal of Semiotics* 39, nos. 3–4: 205–16.

De Man, Paul. 1979. "Autobiography as De-facement." *Modern Language Notes* 94: 919–30.

Dix, Hywel. 2022. "Autofiction." *Oxford Research Encyclopedias*. March 23. https://oxfordre.com/literature/view/10.1093/acrefore/9780190201098.001.0001/acrefore-9780190201098-e-1339.

Dreiser, Theodore, 1998 [originally 1900]. *Sister Carrie*. Philadelphia: University of Pennsylvania Press.

Eagleton, Terry. 1983. *Literary Theory: An Introduction*. Oxford: Basil Blackwell.

Eakin, Paul John. 2020. *Narrative, History, Autobiography*. New York: Routledge.

Edgerton, Robert B. 2000. "Traditional Beliefs and Practices – Are Some Better than Others?" In *Culture Matters. How Values Shape Human Progress*, ed. Lawrence E. Harrison and Samuel Press, 126–40. Huntington. New York: Basic Books.

Foley, Barbara. 1986. *Telling the Truth. The Theory and Practice of Documentary Fiction*. Ithaca and London: Cornell University Press.

Frye, Joanne. 1986. *Living Stories, Telling Lives. Women and the Novel in Contemporary Experience*. Ann Arbor: University of Michigan Press.

Garson, Marjorie. 2000. "Alice Munro and Charlotte Brontë." *University of Toronto Quarterly* 69, no. 4: 783–825.

Gault, Cinda. 2006. "The Two Addies: Maternity and Language in Alice Munro's *Lives of Girls and Women* and William Faulkner's *As I Lay Dying*." *American Review of Canadian Studies* 36, no. 3: 440–57.

Giordani, Maria Rosa. 1995. "Gustave Flaubert, Alice Munro: Due microcosmi a confronto." *Rivista de Studi Canadesi* 8: 187–94.

Gusdorf, George. 1980. "Conditions and Limits of Autobiography." In *Autobiography: Essays Theoretical and Critical*, ed. James Olney, 28–48. Princeton: Princeton University Press.

Häsner, Bernd. 2001. *Metalepsen: Zur Genese, Systematik und Funktion transgressiver Erzählweisen.* Berlin: Freie Universität.

Hornung, Alfred, ed. 2010. "Auto/Biography and Mediation: Introduction." In *Auto/Biography and Mediation*, xi–xviii. Heidelberg: Winter.

Howarth, William. 1980. "Some Principles of Autobiography." In *Autobiography: Essays Theoretical and Critical*, ed. James Olney, 84–114. Princeton: Princeton University Press.

Howells, Coral Ann. 2017. "Writing by Women." In *The Cambridge Companion to Canadian Literature*, ed. Eva-Marie Kröller, 217–38. Cambridge: Cambridge University Press.

Howes, Craig. 2020. "Life Writing." *Oxford Research Encyclopedias*. October 27. https://oxfordre.com/literature/view/10.1093/acrefore/9780190201098.001.0001/acrefore-9780190201098-e-1146?rskey=rYaqTH&result=1#acrefore-9780190201098-e-1146-div1-6.

Hüsgen, Candela, Nienke Peters-Scheffer, and Robert Didden. 2022. "A Systematic Review of Dog-Assisted Therapy in Children with Behavioural and Developmental Disorders." *Advances in Neurodevelopmental Disorders* 6: 1–10. https://link.springer.com/article/10.1007/s41252-022-00239-9.

Hughes, Robert. 2013. "John Steinbeck's Shifting View of America: From *Travels with Charley* to *America and Americans*." *A Political Companion to John Steinbeck*, ed. Cyrus Ernesto Zirakzadeh and Simon Stow, 311–24. Lexington: University Press of Kentucky.

Jolly, Margaretta, ed. 2001. *Encyclopedia of Life Writing: Autobiographical and Biographical Forms*. London and Chicago: Fitzroy Dearborn.

Kendall, Paul Murray. 2022. "Biography." *Encyclopaedia Britannica*. www.britannica.com/art/biography-narrative-genre/Memoirs-and-reminiscences.

Kuiper, Kathleen. 2022. "Sister Carrie." *Encyclopaedia Britannica*. May 25. www.britannica.com/topic/Sister-Carrie.

Malik, Fatima, and Raman Marwaha. 2022. "Developmental Stages of Social Emotional Development in Children."*National Library of Medicine. National Center for Biotechnology Information*. February 7. https://www.ncbi.nlm.nih.gov/books/NBK534819/.

Marra, Giulio. 1996. "Recovering the Past and Story Structure in Margaret Atwood's 'Hack Wednesday' and Alice Munro's 'Age of Faith.'" *Rivista de Studi Canadesi* 9: 73–86.

Martin, Walter. 1979. "Alice Munro and James Joyce." *Journal of Canadian Fiction* 24: 26.

McGrath, Charles. 2012. "The Sense of an Ending." *The New York Times*, November 16.

McWilliams, Eileen. 2007. "Transatlantic Communities: The Joycian Influence in Alice Munro's *Lives of Girls and Women*." In *Narratives of Community: Women's Short Story Sequences*, ed. Roxanne Harde, 372–88. Newcastle: Cambridge Scholars.

Mineo, Liz. 2017. "Good Genes Are Nice, But Joy Is Better." *Harvard University Gazette* online edition, April 11. https://news.harvard.edu/gazette/story/2017/04/over-nearly-80-years-harvard-study-has-been-showing-how-to-live-a-healthy-and-happy-life/.

Montgomery, Lucy Maud. 2013 [originally 1908]. *Anne of Green Gables*. London: Vintage Children's Classics.

Munro, Alice. 2013. "'The Eye,' 'Night,' 'Voices,' 'Dear Life.'" In *Dear Life*, ed. Alice Munro, 257–319. New York: Vintage Books.

Nichols, Bill, 1981. *Ideology and the Image: Social Representation in the Cinema and Other Media*. Bloomington: Indiana University Press.

Österling, Anders. 1969. "Presentation Speech on John Steinbeck." In *Nobel Lectures, Literature 1901–1967*, ed. Horst Frenz, n. pag. Amsterdam: Elsevier. www.nobelprize.org/prizes/literature/1962/ceremony-speech/.

Oliver, Mary Beth, and Julia Woolley. 2011. "Tragic and Poignant Entertainment: The Gratifications of Meaningfulness as Emotional Response." In *The Routledge Handbook of Emotions and Mass Media*, ed. Katrin Döveling, Christian von Scheve, and Elly A. Konjin, 134–47. New York: Routledge.

Olney, James. 2017 [originally 1972]. *Metaphors of Self: The Meaning of Autobiography*. Princeton: Princeton University Press.

Parini, Jay. 2000. "Introduction to *Travels with Charley in Search of America*." In *Travels with Charley in Search of America*, ed. John Steinbeck, vii–xxii. London: Penguin.

Ríos, Diana Isabel Arredondo, and Mari Castañeda. 2011. *Soap Operas and Telenovelas in the Digital Age: Global Industries and New Audiences*. New York: Lang.

Schultz, Jeffrey, and Luchen Li. 2005. *Critical Companion to John Steinbeck: A Literary Reference to His Life and Work*. New York: Checkmark Books.

Serafin, Steven. 2022. "Alice Munro." *Encyclopaedia Britannica*. July 5. www.britannica.com/biography/Alice-Munro.

Smith, Sidonie. 2019. "Autobiographical Inscription and the Identity Assemblage." In *Inscribed Identities: Life Writing as Self-Realization*, ed. Joan Ramon Resina, 75–90. New York: Routledge.

Statistics Canada. 2019. *A Portrait of Canadian Youth*. www150.statcan.gc.ca/n1/pub/11-627-m/11-627-m2019046-eng.htm.

Steinbeck, John. 1962. "Banquet Speech." *Nobel Prize Acceptance Speech*. www.nobelprize.org/prizes/literature/1962/steinbeck/speech/.

Steinbeck, John. 1975. *Steinbeck: A Life in Letters*. Ed. Elaine Steinbeck and Robert Wallsten. New York: Viking Penguin.

Steinbeck, John. 2000 [originally 1961]. *Travels with Charley in Search of America*. London: Penguin.

Stich, Klaus. 1989. "The Cather Connection and Alice Munro's 'Dulse.'" *Modern Language Studies* 19, no. 4: 102–11.

Straubhaar, Joseph Dean, and Antonio La Pastina. 2005. "Multiple Proximities between Television Genres and Audiences: The Schism between Telenovelas' Global Distribution and Local Consumption." *Gazette* 67, no. 3: 271–88.

Thacker, Robert. 2022. "Alice Munro: Biographical." Nobel Prize Outreach AB. www.nobelprize.org/prizes/literature/2013/munro/biographical.

Thacker, Robert. 1992. "Alice Munro's Willa Cather." *Canadian Literature* 134: 42–57.

The Steinbeck Institute at Stanford University. 2022. https://steinbeck.stanford.edu/writer.

Thoreau, Henry David. 1854. *Walden*. Boston: Ticknor and Fields [Norton Critical Edition 2008, ed. William Rossi. New York/London: W.W. Norton & Company].

Wall, Kathleen. 1997. "Representing the Other Body. Frame Narratives in Margaret Atwood's 'Giving Birth' and Alice Munro's 'Meneseteung.'" *Canadian Literature* 154: 74–90.

Wilde, Oscar. 1891. "The Decay of Lying – An Observation." *Intentions. The Project Gutenberg eBook*. https://www.gutenberg.org/files/887/887-h/887-h.htm.

Sofie Behluli

Lonely Descriptions of Art and Life: Olivia Laing's *The Lonely City* (2016)

This article examines how contemporary life writing uses ekphrastic descriptions of visual art to expand personal accounts of life into shared life narratives. By evoking real artworks – public contact zones where numerous lives meet, merge, and create something new – these narratives establish a shared community of images and invite us to rethink individual lives as part of a broader network. In line with the recent interest in the intersection of life writing, intermediality, and American studies (Balestrini and Bergmann 2018), this article discusses entangled life narratives that are both entrenched in American culture yet yearn to surpass it via the confluence of the verbal and visual. The primary focus will be on Olivia Laing's memoir *The Lonely City* (2016), which shifts the author's isolating experience of loneliness into the public sphere by staging vivid, affective, and deeply personal encounters with visual chronicles of solitude.

In the first sentence of her memoir, *The Lonely City*, Olivia Laing verbalizes a request that implicitly underlies her entire book: she asks her readers to imagine a scene and to embody it.

> IMAGINE STANDING BY A WINDOW at night, on the sixth or seventeenth or forty-third floor of a building. The city reveals itself as a set of cells, a hundred thousand windows, some darkened and some flooded with green or white or golden light. Inside, strangers swim to and fro, attending to the business of their private hours. You can see them, but you can't reach them, and so this commonplace urban phenomenon, available in any city of the world on any night, conveys to even the most social a tremor of loneliness, its uneasy combination of separation and exposure. (Laing 2016, 3)

Readers are compelled to picture "standing by a window at night," looking out into a big city, seeking to connect with strangers. The combined metaphors of the prison ("set of cells") and the fishbowl ("strangers swim to and fro") evoke an urban setting in which people – including the lonely author herself – are isolated and exposed. By creating this vivid scene of a hundred thousand illuminated fishbowl cells, *The Lonely City* immediately indicates its interest in sight, imagery, and in visuality. This beginning also introduces a literary strategy which, with the help of visual descriptions, seeks to connect personal experiences with social implications. In fact, Laing's definition of loneliness as an "uneasy combination of separation and exposure" is also a fitting way to describe descriptions: descriptions, to put it simply, aim to summon an absent object for the imagination to behold. Similar to loneliness, they are based on a relationship between presence

https://doi.org/10.1515/9783111551968-008

and absence, proximity and distance. In the case of loneliness, this relationship is social; in the case of description, it is mimetic.

This article examines how Laing uses visual descriptions – especially ekphrases of famous art produced in and influenced by New York City – to challenge the notion of loneliness and to expand her memoir into a shared life narrative. By evoking real artworks by U.S.-American artists such as Edward Hopper and Zoe Leonard, in other words, by making her readers see what she has seen, Laing creates contact zones of the imagination, where numerous lives meet, merge, and produce something new. Similar to other life narratives that strategically describe visual art, Laing's book establishes a shared community of images and invites readers to rethink individual lives as part of a broader network.

Laing uses a host of strategies to achieve these relational effects, but this article focuses primarily on her use of 'ekphrasis.' Simply put, ekphrases are "verbal representations of visual representations" (Heffernan 1993, 3). In their most narrow sense ekphrases are verbal "encounters" with visual art (Kennedy 2012, 31) that invite readers to "'see' the subject in their mind's eye" (Webb 2009, 2). They are vivid, detail-oriented descriptions that turn readers into viewers who can behold artworks, maps, and other visual representations in their imagination. Since an ekphrasis removes reality by two layers of mediation, it draws our attention to the medium itself and forces us to think about the mechanics of writing. Ekphrasis is both illusion and a peek behind the curtain of illusion. By formalizing the point of view or gaze of the speaking subject so effectively, ekphrasis lends itself particularly well to the practice and study of life writing. The above-quoted passage from *The Lonely City* is not an ekphrasis in this narrow sense, as it does not represent something that already represents something else, but it includes *traces* of ekphrases: as I discuss further below, the metaphors of the city as a prison and fishbowl are direct references to Edward Hopper's *Nighthawks* (1942; Figure 1), whereas the use of the word "stranger" foreshadows Laing's ekphrases of Zoe Leonard's *Strange Fruit* (1992–1997; Figure 2). These subtle hints to Hopper's and Leonard's artworks appear as explicit intertextual references only upon rereading the book.[1]

One can observe a similar confluence of the verbal and visual to evoke shared life narratives in various other contemporary works of life writing. Of course, the most obvious example is the rich genre of the artist biography, but then there are also more experimental texts such as Laura Cumming's *The Van-*

1 There are numerous other Hopper-inspired books, for example, Anne Carson's *Men in the Off Hours* (2000), and the short story collections *In Sunlight or in Shadow* (2016, ed. Lawrence Block) and *Pictures from Hopper* (2011, Neil Campbell).

ishing Man (2016), Rachel Cusk's *The Last Supper* (2009) and Michael White's *Travels in Vermeer* (2015). Suffice it to say that Laing's memoir – as a disciplinary and generic hybrid text with a relational approach to depicting lives and a creative use of intermedial strategies such as ekphrasis – is a representative text of a broader contemporary phenomenon.

The link between life writing and visual art has been intuitive for centuries. It partially builds on the age-old comparison of the sister arts of writing and painting and proclaims formulas such as "biography is to portraiture as autobiography is to self-portraiture" (Adams 2000, 226).[2] Considering the history of literature alone, one can think of numerous examples that creatively employ text and image within the context of life writing: Oscar Wilde's *The Picture of Dorian Gray* (1890) traces the fictional life of a protagonist, whose progressing moral corruption is reflected in the titular portrait of himself; Virginia Woolf's satirical *Orlando: A Biography* (1928) features photographs of Woolf's friend and lover, Vita Sackville-West, who poses as the titular and fictional character, Orlando; Ernest Hemingway's memoir *A Moveable Feast* (1924) is supplemented by a variety of photographs; Art Spiegelman's non-fictional graphic novel *Maus: A Survivor's Tale* (1991) relies on both drawings and text to convey his father's experiences as a Polish Jew and Holocaust survivor; and Tracy Chevalier's *Girl With a Pearl Earring* (1999), to name just one more example, fictionalizes the life of the model from Johannes Vermeer's eponymous painting. Whether they are fictional, non-fictional or a hybrid of fact and fiction, these works combine word and image to capture the complexities of life.

By paying attention to the visual arts, this study is also aligned with a growing trend in life writing studies to foreground the medial and material basis of these texts. As Gabriele Rippl notes, "[a]utobiograpy and life-writing studies have recently undergone . . . an intermedial turn" (2018, 147). This intermedial emphasis is particularly effective in understanding life narratives that are interested in connecting different lives across time and space. Such collective, shared or relational life narratives benefit from media-specific analyses since the occurrence, evocation or reference to a medium different than text always already indicates otherness: text and image, self and other, my life and their life. Here is how Nassim Winnie Balestrini and Ina Bergmann put it in their volume on *Intermediality, Life Writing and American Studies*: "research in relational life writing benefits from dissecting discursive features that are rooted in medium-specific character-

2 Adams is clearly modifying classical dictums such as Simonides of Chios's *poesia tacens, picture loquens* [painting is mute poetry, poetry a talking picture] and Horace's *ut pictura poesis* [as in painting so in poetry].

istics. Such subtlety of multiple voices has contributed to expanding the generic range of life narratives, particularly through media combinations and intermedial imitation of medium-specific sign systems" (2018, 2). Clearly, the study of textual references to visual art lends itself to the study of shared life narratives. Both are based on the act of translation: Ekphrasis translates the visual medium into the verbal one, whereas life writing translates lived, embodied experiences into a textual archive of life.

Balestrini and Bergmann's link of media combination with the notion of 'otherness' goes back to W. J. T. Mitchell's seminal 1994 book *Picture Theory*. Talking specifically about the uses of ekphrases, Mitchell argues:

> The 'otherness' of visual representation from the standpoint of textuality may be anything from a professional competition (the *paragone* of poet and painter) to a relation of political, disciplinary, or cultural domination in which the 'self' is understood to be an active, speaking, seeing subject, while the 'other' is projected as a passive, seen, and (usually) silent object. (1994, 157)

Ekphrasis, to follow Mitchell's argument, stages not only an encounter with another, a visual medium, but also with a foreignness or otherness to which we can ascribe power relations. The here mentioned "passive, seen and (usually) silent object" can stand for women, colonized subjects, and other marginalized or oppressed groups. A media-specific analysis, in other words, is not only a potent approach to polyvocal and genre-contesting texts, but also a method to unravel the power hierarchies within them. This is especially pertinent in the context of life writing, which often tends to examine the broader structures that exert power on the individual – be they familial, institutional, or political, to name just a few.

This article aims to examine a creative landscape in which life writing and visual art converge. It asks how representations of visual art feature into life narratives, what aesthetic functions they fulfill, and how they impact the generic borders of the genre. I approach these questions by offering some preliminary remarks on life writing in a first step and by discussing Olivia Laing's memoir more closely in a second.

Preliminary remarks on life writing

Zachary Leader informs us that 'life writing' is "a generic term used to describe a range of writings about lives or parts of lives, or which provide materials out of which lives or parts of lives are composed" (2015, 1; cf. Rippl, Schweighauser, and Steffen 2013, 7). The term 'life writing,' which is said to have been coined by Virginia Woolf in her essay "A Sketch of the Past" (Woolf 1939, 80), subsumes a vast variety of texts including autobiography, biography, memoirs, diaries, letters, and more experimental subgenres such as autofiction, biofiction, creative non-fiction, and auto/biografiction. Scholars such as Julia Novak have noted a rising interest in life writing in the past few decades, using catchphrases such as 'biography boom' and 'memoir craze' (2017, 1). Indeed, a large section of today's literary market is dominated by writers who fall into or are adjacent to the life writing genre, including Maggie Nelson, Karl Ove Knausegaard, and Rachel Cusk. They all build on a long tradition of writing that has been established and popularized by writers such as St Augustine, Jean-Jacques Rousseau, Benjamin Franklin, Anne Frank, and many others. "The present avalanche of autobiographical writing," state Kerstin Shands and the co-editors of *Writing the Self*, "indicates that our time has a great interest in questions of self and identity" (2015, 7). With the proliferation of smartphones, social media, and digital media more broadly, the interest in notions of the self has become more prevalent than ever. Taking and posting a selfie is a form of direct participation in this life writing culture, a way of fashioning one's public image, or a form of simultaneous living and recording life.

The same time late twentieth century conditions that gave rise to this boom in life writing also challenged the validity of some of the fundamental pillars on which life writing is built: authenticity, truth, representation, and, most dramatically, the self. As poststructuralist thinkers contested the very basis of language, life writing practitioners and theorists saw themselves faced with a new host of problems. The line between fact and fiction was blurring and the 'truth'-based foundation of life writing was eroding. All narratives – whether they are fictional novels or autobiographies – can and must be placed on a spectrum from factual to fictional. Indeed, John Paul Eakin states as early as in 1985 that the self as "the center of all autobiographical narrative is necessarily a fictive structure" and that all "materials of the past are shaped by memory and imagination to serve the needs of the present consciousness" (Eakin 1985, 3, 5–6). In this new understanding of life writing, creativity and imagination are stable components of all life narratives.

This softening of the fact/fiction divide has not only birthed new genres such as autofiction, it has also shown that "the distinction between self and other is not always clear" (Leader 2015, 2). This unclear border between self and other is crucial for life narratives that have relational qualities, i.e. for shared life narratives that emphasize the entanglements of different lives. This relational expansion of the self further multiplies an already pluralistic conception of 'self' in life writing, adding to the split between the 'experiencing self' that is at the center of the life narrative and the 'writing self' that looks back on the experiencing self with a temporal distance.[3] Relationality in life writing asks us to understand the self not only as something that contains multitudes, but also as something that flows into, intersects with, and is shaped by others.

At the back of their book on *Reading Autobiography* (2000), Sidonie Smith and Julia Watson have compiled a toolkit with which to approach autobiographical texts and other life narratives. Amongst the plethora of research questions they propose, they also feature a section on 'relationality': "What others inhabit the text? Is there a significant other posited in the text through whom, to whom, or about whom the life narrative is narrated? . . . . What is the impact of this *relationality* upon your understanding of the rhetorical 'I' or the narrator's subjectivity?" (Smith and Watson 2000, 59). These and other, previously-posed questions guide my reading of Laing's visual and relational memoir below.[4]

Reading and Seeing *The Lonely City*

Throughout her career, Olivia Laing has been drawn to life writing with all its upsides and downsides. Born in 1977 in Buckinghamshire, England, the writer and critic first published *To the River: A Journey Beneath the Surface* (2011), in which she connects the lives of authors such as Virginia Woolf to rivers and reflections on life writing. This blend of "biography, history, nature writing and memoir," as the book is described on Laing's official website, is followed by *The Trip to Echo Spring: On Writers and Drinking* (2017a). Like her first book, *The Trip to Echo Spring* interweaves biographical writing with memoir conventions, all the while reflecting on the broader societal, historical, and psychological parameters of alcoholism. In her third book, *The Lonely City* (2016), Laing explores her own experiences of loneliness in New York City at a particular time in her life and she connects those experiences to various New York-based artists and their visual art.

After three books on rivers, alcoholism, and loneliness, all of which shed light on the writer's own life by inspecting the lives of others and, crucially, vice versa,

3 Critics have identified many more autobiographical 'selves' than the 'experiencing self' and the 'writing self' (cf. Smith and Watson 2000, 59).

4 A memoir distinguishes itself from the more traditional autobiography by prioritizing themes and topics in an enclosed period of life. As Marcus Billson explains it, autobiographies focus on "becoming in the world," whereas memoirs are dedicated to "being in the world" (qtd. in Buss 2001, 595). Memoirs also tend to "personalize history and historicize the personal" (595). In recent decades memoirs have become more popular, perhaps because they are allowed to intermingle memory with imagination, personal experience with history, and the self with the other a bit more freely (cf. Marcus 2018, 6–7).

Olivia Laing published *Crudo* in 2017, an autofictional novel set in the tumultuous years of Donald Trump's presidency; *Crudo* was followed by *Funny Weather: Art in an Emergency* (2020), a collection that brings together more than 50 essays on art and culture that Laing had previously published throughout her career; and in 2021 she published *Everybody: A Book about Freedom*, a book that reflects on conceptions of 'body,' its political relations to freedom and, of course, on her own embodied experiences as a queer, non-binary person. This survey of her published work makes three things clear: First, Olivia Laing is interested in grand themes, which allow her to bring various disciplines, genres, and multiple voices across time and space in dialogue with each other; second, the practices of writing and creating art and the lives of writers and artists have always been a preoccupation in her own work; and third, Laing is deeply invested in making sense of her own life and craft via the life and art of others, just as much as she is in making sense of the lives of others through her own experiences. Her writing, which can be described as 'creative non-fiction,' is inherently relational and transgressive.

The Lonely City: Adventures in the Art of Being Alone is a memoir on loneliness that also sketches the biographies of numerous New York-based artists and how they explored loneliness in their art. The book records a year or so in Olivia Laing's life in her mid-thirties and tells of her experiences of loneliness when, soon after moving from the UK to New York City for her then-boyfriend, he breaks up with her. In addition to this romantic breakup, then, Laing is also isolated through this transatlantic move, which establishes her as a foreign subject who casts an outsider's gaze onto the United States. Thus fully isolated, loneliness builds the core experience that underlies the formal, thematic, and aesthetic project of Laing's memoir. Indeed, the memoir is a creative response to this personal crisis. As Laing states early on,

> Loneliness is difficult to confess; difficult too to categorise. Like depression, a state with which it often intersects, it can run deep in the fabric of a person, as much a part of one's being as laughing easily or having red hair. Then again, it can be transient, lapping in and out in reaction to external circumstance, like the loneliness that follows on the heels of a bereavement, break-up or change in social circles. (2016, 4)

In this paragraph, readers encounter the phrase that "[l]oneliness is difficult to confess," which not only sets the stakes of this book but also aligns it with a tradition of other confessions – including those of St Augustine and Jean-Jacques Rousseau. While loneliness can be considered "as much a part of one's being as laughing easily or having red hair," there is yet something shameful and isolating about it; something that makes it an uncomfortable topic to discuss with others. We as readers may ask ourselves why she mentions the word "confession" early in the text. When Laing goes on to connect loneliness to social circumstances such as “bereavement, break-up or change in social circles,” rather than classifying it merely as a personal attribute, loneliness loses its sharp edge. In sum, this early paragraph asks whether loneliness is a matter of the self or one of society. By asking this question, Laing’s memoir also displays a tendency to add critical reflections on top of the life narrative: this book writes life *and* reflects on the practices of writing lives on a metalevel.

The introduction goes on to pose a host of questions that Laing engages with throughout her book:

> There were things that burned away at me, not only as a private individual, but also as a citizen of our century, our pixelated age. What does it mean to be lonely? How do we live, if we’re not intimately engaged with another human being? . . . And is technology helping with these things? Does it draw us closer together, or trap us behind screens? (2016, 5)

Laing’s guiding questions reveal that she is interested in everything from the minute experiences of private life to large-scale connections between history, technology, and psychology. As the English author further explains, she plans to find some answers to these questions by focusing on visual art, or, more precisely, visual art created in New York City over the past seventy or so years. Amongst the plethora of artists she discusses throughout her book, Laing focuses on four “documenters of the lonely city” (Laing 2016, 6): Edward Hopper, Andy Warhol, Henry Darger, and David Wojnarowicz. Their photographic portraits are reproduced in Laing’s book, highlighting their alterity through media combination. The reproductions are always given at the beginning of a new chapter, signaling the theme of the coming section. Their medial otherness already conveys a multiplicity of voices that the book will grapple with, aiming to interweave the biographies of these men with their ekphrastically-evoked art and, of course, with Olivia Laing’s own experiences of loneliness. It should be noted at this point that these four portraits and five other photographs are reproduced in the book, whereas the ekphrastically evoked artworks of the artists are not. In other words, Laing gives us photographic portraits of the artists – the other subjects of this relational life narrative – but the artworks she discusses are merely present through her descriptions. Readers can easily look them up on the Internet, if they so choose, or they can simply follow Laing’s descriptions and re-inscriptions.

Early on in her memoir, Laing also introduces the metaphor of the map in order to connect the city with its citizens and the notion of ‘loneliness.’ Indeed, it is as if she is trying to spatialize ‘loneliness’ into a map to better grasp and represent this complex experience: “What I was building in those years, and what now follows, is a map of loneliness, built out of both need and interest, pieced together from my own experiences and those of others” (Laing 2016, 8). The idea of the map allows Laing to place herself into the same imaginary space as Warhol, Hopper, Darger, and Wojnarowicz, not only to self-canonize her own art as being a worthy addition to this group of creators, but also to evoke a relational subjectivity through which loneliness can be processed and perhaps overcome.

Due to reasons of scope, I cannot discuss all ekphrastic evocations in Laing's book but will focus on a few exemplary ones from the beginning and end. By omitting the ekphrastic descriptions in the book's middle section and only discussing those at the beginning and end, I show how Laing's emotional journey from loneliness to feeling more connected with others is reflected in the artworks she discusses. This emotional journey begins with reflections on Edward Hopper's work, onto which Laing projects her loneliness, and ends with a discussion of Zoe Leonard's work, in which Laing sees a delicate attempt to stitch the self back together after the experience of loss and to connect with others.

Laing first examines Edward Hopper's work in the second chapter, which is titled "Walls of Glass." Edward Hopper (1882–1967) is one of the most renowned American artists of the twentieth century and is best known as a chronicler of American urban solitude. As previously mentioned, Laing has just arrived in New York City and been left by the man for whom she moved across the Atlantic:

> My life felt empty and unreal and I was embarrassed about its thinness, the way one might be embarrassed about wearing a stained or threadbare piece of clothing. . . . If I could have put what I was feeling into words, the words would have been an infant's wail*: I don't want to be alone. I want someone to want me. I'm lonely. I'm scared. I need to be loved, to be touched, to be held.* (Laing 2016, 14)

True to the previously mentioned 'confessions,' these lines read like the most personal and vulnerable expressions of interiority that a person can express. One can also observe the split between the experiencing self and the writing self, as Laing has found the right words to communicate her past experiences only now that she is looking back at her life. When she continues these intimate observations of herself, she brings in visual art for the first time, comparing herself to the subjects in Hopper's paintings. Here, she evokes two paintings called *Automat* (1927) and *Morning Sun* (1952), whose iconic status and immediate recognizability do a lot of the descriptive heavy lifting:

> I knew what I looked like. I looked like a woman in a Hopper painting. The girl in *Automat*, maybe, in a cloche hat and green coat, gazing into a cup of coffee, the window behind her reflecting two rows of lights, swimming into blackness. Or the one in *Morning Sun*, who sits on her bed, hair twisted into a messy bun, gazing through her window at the city beyond. . . . I often sat just like that, adrift in rumpled sheets, trying not to feel, trying simply to take consecutive breaths. (Laing 2016, 15–16)

Laing embodies and identifies with these two female figures from Hopper's oeuvre, both of whom are represented as solitary, isolated individuals who only have access to the city around them via natural or artificial light. Hopper's women are embedded in a world of cold colors, where light and dark hint at an inner turmoil that is further highlighted by their vulnerable body postures. Both women are cut off from the world, entrapped by an architecture of glass and oil paint. This is, at least, how Laing reads these paintings and how she projects her own feelings onto them. Later on in the book, Laing not only identifies with Hopper's painted figures but also with the artist himself: "perhaps it was Edward Hopper I was morphing into. Like him, I found myself becoming a peeper, a creeper, a connoisseur of open windows, patrolling in search of stimulating sights. Like him, my attention was often caught by the erotic" (Laing 2016, 223).

It makes sense that Laing relates both to the artist and to his painted women, given that she is both the creator and the subject of her own memoir. Laing observes and is being observed; she represents and is being represented. The peeping, creeping, and invasive nature of the endeavor that she mentions in the quote is also fitting of the genre, because, as we know from Hermione Lee, "we always greedily want moments of intimacy, revelation, immediacy, and inwardness" (2015, 125). Indeed, memoirs and life writing in general always have a voyeuristic aspect to them.

Figure 1: Edward Hopper, *Nighthawks*, oil on canvas, 1942, The Art Institute of Chicago, © Heirs of Josephine N. Hopper / 2024, ProLitteris, Zurich.

Laing also presents her readers with a three-page long ekphrasis of Edward Hopper's most famous work, *Nighthawks* (1942; Figure 1). Inspired by the 1940s streets of New York City and yet, eerily unclear about its specific time and place, this painting is often claimed to perfectly encapsulate the experience of modern

Americans. Depicting an all-night diner in which three customers all seem to be lost in their own thoughts, this painting is the ultimate representation of the close-but-far, together-but-alone feeling that Olivia Laing mentions at the beginning of her book. Here is how she describes *Nighthawks*:

> I don't suppose there are many people in the western world who haven't peered into the cool green icebox of that painting. . . . The diner was a place of refuge, absolutely, but there was no visible entrance, no way to get in or out. There was a cartoonish, ochre-coloured door at the back of the painting, leading perhaps into a grimy kitchen. But from the street, the room was sealed: an urban aquarium, a glass cell.
>
> Inside, in their livid yellow prison, were the four famous figures. . . . Green on green, glass on glass, a mood that expanded the longer I lingered, breeding disquiet. . . . It was impossible to gaze through into the diner's luminous interior without experiencing a swift apprehension of loneliness, of how it might feel to be shut out, standing alone in the cooling air. (Laing 2016, 20–23)

Alluding to the fact that most people have seen this famous painting at one point or another, Olivia Laing immediately establishes a connection between herself and others. This canonical work of art acts as a meeting point, as an imaginary site where the author, the artist, and the reader can enter into dialogue with each other. Olivia Laing pays attention to the painting's visual semantics, including its color and composition, and she also creatively interprets it as "an urban aquarium" and a "glass cell." Both of these metaphors tie in with the book's opening lines, in which the city is represented as a prison and its inhabitants as prisoners. The idea of the 'prison' is also explicitly repeated here. The last line from this scene, however, is rather surprising as it shows that Olivia Laing imagines herself *outside* of this prison, "shut out [and] standing alone in the cooling air," rather than inside of it. Even though Hopper's figures are trapped in their own way, unable to connect to the people that are in the urban aquarium with them, Laing yearns to be with them. This is a more implicit strategy to communicate just how lonely she must have felt at the time.

After this chapter on Hopper, Laing uses the adjective 'Hopperesque' to refer to other artists' use of color, perspective, and theme, but, apart from that, Edward Hopper does not feature prominently in the rest of the book. Laing leaves Hopper and his art behind as the book traces her own development from deepest despair and loneliness to a newfound sense of self, one connected to others. This is not to say that Hopper's art is inherently "lonelier" than Leonard's art, but only that Laing imbues a stronger sense of loneliness in the former than the latter. After all, Hopper's and Leonard's artworks are multifaceted pieces that are open to a variety of interpretations, so the art critical readings that are given in *The Lonely City* tell us more about Laing – the seeing, feeling, and interpreting agent of this life narrative – than the artworks themselves.

My analysis makes a huge jump now to discuss the ekphrases at the end of Laing's memoir, thereby omitting her contemplations of Andy Warhol, Peter Hujar, Jean-Michel Basquiat, and other influential New York artists. Without going too much into detail here, Laing traces their biographies, professional interactions, friendships, and shared grief over the AIDS epidemic. By diving into this pool of artists, who were connected to each other in various personal and artistic ways, Laing draws from an already established network of people for her relational, collective life narrative. What most of them have in common is their direct or indirect contact with AIDS, and the stigmatization and marginalization they experienced as a consequence of being part of this queer community. By turning her gaze to this group of othered, ostracized individuals – not merely an elite group of artists –, Laing forces readers to think more intently about the subjects of life writing: Which lives are being recorded and which ones are overlooked? Which lives are, ultimately, worth living and telling? These might seem like odd questions to pose in the context of a memoir about a white, western, middle-class woman, but it is clear that Laing does not necessarily view herself just as part of this privileged group. Indeed, she strongly identifies with this collection of queer men and queerness, more broadly, runs through the entire fabric of the memoir. Only after *The Lonely City* was published did Laing publicly come out as a queer, non-binary person, so her personal investment in the lives of Hujar, Wojnarowicz, and the others mostly remains unspoken. The closest she ever comes to making this connection between her life and their lives explicit is when she states:

> I'd never been comfortable with the demands of femininity, had always felt more like a boy, a gay boy, that I inhabited a gender position somewhere between the binaries of male and female, some impossible other, some impossible both. Trans, I was starting to realise, which isn't to say I was transitioning from one thing to another, but rather that I inhabited a space in the centre, which didn't exist, except there I was. (Laing 2016, 125)

Suddenly, the loneliness experienced from having lost a romantic connection and being a foreigner on a different continent is multiplied by this layer of non-conforming identity. To say that one inhabits "a space in the centre, which didn't exist" is to say that one feels invisible, excluded, and erased. In an interview for *The Sydney Morning Herald*, Laing expresses regret for not having "been more emphatic in *The Lonely City* about identifying as queer" (Kembrey 2018, n. pag.).

While Laing's internal processes and affective connections are not always brought to the fore of the text, it is clear that she finds solace, peace, and community in her engagement with this group of New York-artists. This is also why the book culminates with an ekphrastic examination of Zoe Leonard's installation *Strange Fruit* (1992–1997; Figure 2), which is an artistic response to the lives lost to AIDS.[5] In spite of its sad nature, Laing projects feelings of hope onto Leonard's artwork and therefore communicates her own progress in overcoming loneliness and despair.

Figure 2: Detail from Zoe Leonard, *Strange Fruit*, fruit, thread and other media, 1992–1997, Philadelphia Museum of Art, © Zoe Leonard; Courtesy the artist, Galerie Gisela Capitain, Cologne and Hauser & Wirth.

Zoe Leonard's *Strange Fruit* is discussed in the final and equally-titled chapter of *The Lonely City*. Leonard (b. 1961) is an American concept artist specialized in photography, sculptures, and installations, and the only female artist to whom Laing pays a significant amount of attention. Owned by the Philadelphia Museum of Art, *Strange Fruit* consists of nearly three hundred fruit skins – bananas, oranges, lemons, and more – that she sewed back up after she ate the fruit. Although it would have been possible to preserve the fruit skins, Leonard decided against that option to better illustrate the ephemeral, decaying nature of life (Quabeck 2019, n. pag.). *Strange Fruit* is an active *memento mori*. Evoking a yearning to be whole again and to defeat the decay of time, this installation is imbued with both sadness and hope. Here is how Olivia Laing describes it close to the end of her book:

5 The title of Leonard's installation references Billie Holiday's 1939 song of the same name, which protests the lynching of African Americans. Leonard recontextualizes the title and expands the group of people who experience marginalization, neglect, and prosecution from African Americans to also encompass the communities affected by AIDS. This also reframes the state's devastating passivity towards the AIDS crisis as a new form of lynching.

> This piece, which is clearly part of the vanitas tradition in art, the practice of depicting matter as it passes from radiance to decomposition, was made in memory of David Wojnarowicz, who had been a close friend of Leonard's. . . . *Strange Fruit* was made during those years of solitude, arising if not in response to then certainly as a consequence of the mass losses of the AIDS years, the exhaustion of labouring to bring about political change. . . . Speechless, *very silent*, the fruit convey in their smallness, their particularity, the pain of breakage, of vanishing, of longing for something beloved that has departed and will not come again. (Laing 2016, 256–61)

Wojnarowicz, the artist in whose memory *Strange Fruit* was created, is one of the central artists that Laing discusses in her book. He passed away from AIDS, just like Peter Hujar and many other New York-based artists who are featured in *The Lonely City*. This ekphrasis thus ties together a few of the themes and lives that the book has developed thus far. And compared to the earlier-discussed ekphrases of Hopper's paintings, which emphasize urban solitude, this ekphrasis of Zoe Leonard's work is all about repairing that which is organic and alive. The sewn-up fruit may be "speechless" and "very silent," but Olivia Laing envoices them with her writing. By emphasizing "their smallness" and "their particularity," and by ascribing a "longing for something beloved that has departed" to them, Laing is also personifying the fruit and turning it into another subject for her relational life narrative. Her ekphrasis continues:

> What makes *Strange Fruit* so deeply touching, so intensely painful, is the work of stitching, which makes legible another aspect of loneliness: its endless agonising hope. Loneliness as a desire for closeness, for joining up, joining in, joining together, for gathering what has otherwise been sundered, abandoned, broken or left in isolation. Loneliness as a longing for integration, for a sense of feeling whole. . . . Art doesn't have to have a reparative function, any more than it has a duty to be beautiful or moral. All the same, there is art that gestures towards repair . . . . (Laing 2016, 262–66)

It is clear from this ekphrasis, which is located at the end of the book, how much the author has progressed emotionally: In contrast to her earlier ekphrases of Hopper's paintings, in which she only saw prisons and loneliness, here she is able to spot an "endless agonizing hope." Her description of *Strange Fruit* gives readers an intimate peek into her current mental state, indicating that the overpowering feeling of loneliness has been softened up by more positive feelings of hope, renewal, and reparation. Similar to the "reparative function" that she mentions here with reference to *Strange Fruit*, this ekphrasis has a reparative function within the book – for the writer as well as the reader.

Indeed, the reader is pulled into narrative at various points and from the moment the memoir opens with the following dedication: "If you're lonely,/this one's for you" (Laing 2016, n. pag.). Reader involvement is also achieved when Laing asks them to imagine a scene or look closer at an artwork, but the final page of the book turns these innocent acts of addressing the reader into a proper plea for solidarity, political activism, and empathy for all humans – especially the marginalized, voiceless, and overlooked. Here is a section from the end of *The Lonely City*, in which Olivia Laing summarizes her findings and puts forward her message:

> I don't believe the cure for loneliness is meeting someone, not necessarily. I think it's about two things: learning how to befriend yourself and understanding that many of the things that seem to afflict us as individuals are in fact a result of larger forces of stigma and exclusion, which can and should be resisted.
>
> Loneliness is personal, and it is also political. Loneliness is collective; it is a city. As to how to inhabit it, there are no rules and nor is there any need to feel shame, only to remember that the pursuit of individual happiness does not trump or excuse our obligations to each another. (Laing 2016, 281)

These sentences attest to Olivia Laing's long journey of improvement, from looking out at the city at night and feeling lonely to realizing that, in many cases, loneliness is a product not of individual behavior but of societal structures of exclusion, shaming, and oppression. These are all feelings that the author grappled with when she was abandoned by the man for whom she moved to New York, at the beginning of the book, but her memoir documents how her engagement with visual art helped her put loneliness into perspective and thus partially overcome it. Loneliness is personal as well as political and collective, Olivia Laing concludes. By reminding her readers that loneliness "is a city," she offers a relational form of life writing and puts multiple lives in connection with each other: she connects her own life with the lives of artists, people who have passed away from AIDS, the real lives behind the painted figures and sewn-up fruit, and, the lives of the readers who have made it to the end of the memoir. These interconnected lives come into sharper focus during the ekphrastic passages of the memoir, as visual art seems to act as a particularly potent site for gazes to meet, for lives to intersect, and for loneliness to be overcome.

Works cited

Adams, Timothy Dow. 2000. *Light Writing & Life Writing: Photography in Autobiography*. Chapel Hill/London: University of North Carolina Press.

Balestrini, Nassim Winnie, and Ina Bergmann. 2018. "Intermediality, Life Writing, and American Studies: A Brief Introduction." In *Intermediality, Life Writing, and American Studies: Interdisciplinary Perspectives*, ed Nassim Winnie Balestrini and Ina Bergmann, 1–8. Boston: de Gruyter.

Buss, Helen M. 2001. "Memoirs." In *Encyclopedia of Life Writing: Autobiographical and Biographical Forms* 2: L-Z, ed Margaretta Jolly, 595–97. Chicago: Fitzroy Dearborn.

Eakin, John Paul. 1985. *Fictions in Autobiography: Studies in the Art of Self-Invention*. Princeton: Princeton University Press.

Heffernan, James A. 1993. *Museum of Words*. Chicago: University of Chicago Press.

Kembrey, Melanie. 2018. "Olivia Laing Interview: Pinning Down the Raw Summer of 2017." *The Sydney Morning Herald*, July 20. https://www.smh.com.au/entertainment/books/olivia-laing-interview-pinning-down-the-raw-summer-of-2017-20180712-h12lag.html.

Kennedy, David. 2012. *The Ekphrastic Encounter in Contemporary British Poetry and Elsewhere*. Farnham: Ashgate.

Laing, Olivia. 2011. *To the River: A Journey Beneath the Surface*. Edinburgh: Canongate.

Laing, Olivia. 2017a. *The Trip to Echo Spring: On Writers and Drinking*. Edinburgh: Canongate. Orig. publ. 2013.

Laing, Olivia. 2017b. *The Lonely City: Adventures in the Art of Being Alone*. Edinburgh: Canongate. Orig. publ. 2016.

Laing, Olivia. 2019. *Crudo*. London: Picador. Orig. publ. 2018.

Laing, Olivia. 2020. *Funny Weather: Art in an Emergency*. London: Picador.

Laing, Olivia. 2021. Official Website. https://www.olivialaing.com.

Leader, Zachary. 2015. "Introduction." In *On Life Writing*, ed. Zachary Leader, 1–6. Oxford: Oxford University Press.

Lee, Hermione. 2015. "'From Memory': Literary Encounters and Life-Writing." In *On Life Writing*, ed. Zachary Leader, 124–41. Oxford: Oxford University Press.

Marcus, Laura. 2018. *Autobiography: A Very Short Introduction*. Oxford: Oxford University Press.

Mitchell, W. J. T. 1994. *Picture Theory: Essays on Verbal and Visual Representation*. Chicago: University of Chicago Press.

Novak, Julia. 2017. "Experiments in Life-Writing: Introduction." In *Experiments in Life-Writing: Intersections of Auto/Biography and Fiction*, ed. Lucia Boldrini and Julia Novak, 1–35. London: Palgrave Macmillan.

Quabeck, Nina. 2019. "Intent in the Making: The Life of Zoe Leonard's 'Strange Fruit.'" *Contemporary Burlington*. May. https://contemporary.burlington.org.uk/journal/journal/intent-in-the-making-the-life-of-zoe-leonards-strange-fruit.

Rippl, Gabriele. 2018. "Ekphrastic Encounters and Word-Photography Configurations in Contemporary Transcultural American Life Writing." In *Intermediality, Life Writing, and American Studies: Interdisciplinary Perspectives*, ed. Nassim Winnie Balestrini and Ina Bergmann, 147–65. Boston: de Gruyter.

Rippl, Gabriele, Philipp Schweighauser, and Therese Steffen. 2013. "Introduction: Life Writing in an Age of Trauma," in *Haunted Narratives: Life Writing in an Age of Trauma*, ed. Gabriele Rippl, Philipp Schweighauser, Tiina Kirss, Margit Sutrop, and Therese Steffen, 3–18. Toronto: University of Toronto Press.

Shands, Kerstin W., Giulia Grillo Mikrut, Dipti R. Pattanaik, and Karen Ferreira-Meyers, eds. 2015. "Introduction." In *Writing the Self: Essays on Autobiography and Autofiction*, 7–28. Södertörns högskola: Elanders.

Smith, Sidonie, and Julia Watson. 2000. *Reading Autobiography: A Guide for Interpreting Life Narratives*. Minneapolis/London: University of Minnesota Press.

Webb, Ruth. 2009. *Ekphrasis, Imagination and Persuasion in Ancient Rhetorical Theory and Practice*. Farnham: Ashgate.

Part III: **The Social**

Atalie Gerhard

Womanist Awakenings in the Multicultural North: How Fatuma Adar's and Méshama Eyob-Austin's Life Writings Represent Black Girlhood in Canada

Introduction

This article analyzes multimedia coming-of-age narratives by Fatuma Adar and Méshama Eyob-Austin that can be situated within the larger genre of life writing. They address Canadian multiculturalism that still includes assimilationist expectations. Although the primary texts were published by a university press and a national news outlet, I compare their aesthetics of countering mainstream denials of racism in Canada by representing personal awakenings to political awareness. Despite differences between Canada and the U.S., I interrogate whether Alice Walker's concept of womanism could describe the advocacy for decolonization and female empowerment that Adar and Eyob-Austin perform. On the one hand, Adar negotiates familial traditions and adolescent self-discovery. On the other hand, Eyob-Austin compares racialized perspectives when she considers her positionality. Both narratives mirror the dynamic of Audre Lorde's *Zami: A New Spelling of My Name* (1982) when they juxtapose desires to experience identity with exclusion therefrom. It will be shown that the narrators derive empowerment from their migrant heritages when they define themselves differently than their surroundings.

Black Canada: between multiculturalism and assimilationism?

"There is no core identity, no mainstream in Canada," remarked Prime Minister Justin Trudeau to *The New York Times* in 2015 (Lawson 2015). However, long before Black Lives Matter protests emerged throughout North America, activists and scholars called for Canada to reckon with its history of anti-Blackness. Indeed, Black Canadian history can be told through the lives of women – a group including slaves, protesters, laborers, migrants, and prisoners but the mainstream denial of their suffering suppresses public awareness of their resistance. In the twenty-first cen-

https://doi.org/10.1515/9783111551968-009

tury, Black women and girls in Canada resist the erasure of their predecessors and express an identity beyond Justin Trudeau's "postnational state" (n. pag.). Notably, Afua Cooper published the first English translation of the narrative that Marie-Joseph Angélique "wrote" when she attempted to escape slavery but was incarcerated, tortured, and executed for allegedly incinerating Old Montreal in April 1734 (Cooper 2006, 6, 12; Clarke 2005, xvii). According to George Elliott Clarke, Angélique was the most famous Black slave in Canada but racism prevented her from becoming a canonized rebel (Clarke 2005, xiv, xv). What impact do the enslavement, incarceration, and hanging of Angélique have for Black Canadians today? On the one hand, Angélique was a freedom fighter because she allegedly destroyed the Old Montreal that white settlers built by occupying Indigenous land and using slave labor (xiv). On the other hand, the history of Black slavery in Canada is often relativized by white historians like Marcel Trudel (Sawallisch 2016, 88, 93–95).

Since the 1960s, Trudel has been identified as an apologist for praising masters for acting as godparents at slave baptisms in Quebec (Sawallisch 2016, 95). Meanwhile, Black U.S.-Americans often idolize Canada as well due to the earlier abolition of slavery by thirty-one years compared to the U.S. in 1865. During this time, up to 40.000 slaves fled into Upper Canada via the Underground Railroad that was conducted by Harriet Tubman who had escaped slavery herself (Henry and McIntosh 2022; Henry 2020). Her life is linked to gospel songs about the guiding "Drinking Gourd" constellation and the mythic northern "Canaan" (Hayes 2012, 80). Such comparisons of post-slavery Canada to the biblical Canaan are rooted in Black identifications with the narrative of Moses leading his people out of Egyptian slavery (Gilroy 1993, 207).

Living alongside whites and participating in settler culture still visualizes integration into the multicultural Canada as defined in 1971 (Thobani 2007, 175, 252). Then-Prime Minister Pierre Trudeau issued his Federal Response supporting multiculturalism and English and French bilingualism in Canada (163). However, Sunera Thobani criticizes Canada for styling itself as a Christian savior that depends upon the dependency of "others" who are suffering (107–09, 114). An extreme example of how violence in Canada was justified by Christian saviorism can be found in the residential school system that confined Indigenous children because "The adoption of Christian names, values, and morals by the children and the inculcation of the work ethic and respect for private property were defined as their means of salvation" (120). Especially welfare services continue to justify intrusions into marginalized families with Christian values while the contributions of non-whites are not equally valued (25, 106–09, 112, 128, 140). Thobani thus problematizes that the stereotypical Canadian is still a white Christian person while people of color are racially profiled, having to "prove" their citizenship to border security (238–42). This increased policing of Blacks and Muslims during the "War on Terror" is echoed by Robyn Maynard (Maynard 2017, 87–88, 114). She challenges the myth of a post-racial Canada by juxtaposing Black incarceration rates with the absence of any increase in criminal behaviors (96, 144–45). Instead, Maynard indicts the excessive use of state force against Black loitering, prostitution, undocumented labor migration, and drug use (4, 46, 48, 91, 96, 141, 144–45, 150, 153, 174). She enumerates prisons, orphanages, and migrant detention centers as exemplary sites of anti-Blackness (6–7).

According to Achille Mbembe, colonialism is indeed directed inwards besides occupation abroad when formerly colonized subjects migrate to First World countries due to hardships at home (Mbembe 2010, 179, 181). Although Mbembe discusses France, his understanding of limited migration, assimilationist imperatives, cyclical marginalization, and public invisibility as shaping the cultural image of nations applies to Canada as well (164, 168, 180, 183). For example, Mbembe invokes colonial history which does not only shape France, but North America as well as a reason for a contemporary racist discrimination that targets especially ghettoized Black Muslim migrant youth under the guise of culturalism (177). As in Thobani's criticism of Canadian Christian saviorism, Mbembe further names both, the desire for border and immigration enforcement and the desire for visibility of the suffering of minorities as examples of the symbolical dispossession of struggling migrants that mirrors the historical subjugation of racial and cultural "others" in France (117). Exclusion throughout society uniquely targets Black women due to their intersecting identity aspects, as pointed out by Kimberle Crenshaw (1989, 140) and Patricia Hill Collins (2000, 5, 12, 127, 146, 281). Albeit in the U.S., they focus on Black women who are denied opportunities at work and in education (Crenshaw 1989, 147, 165; Collins 2000, 4, 16). However, not only racism and sexism are addressed in the life writings from Canada that I discuss in the following. Thus, Méshama Eyob-Austin's and Fatuma Adar's coming-of-age narratives hold the mirror to exclusionary myths of the multicultural North.

Discourses like the ones their narratives participate in are as crucial as they are rare in Canada at the beginning of the twenty-first century. Yet, in light of the contemporary cooptation of diversity by liberal mainstream politics, Adar's and Eyob-Austin's narratives face two obstacles in order to voice holistic criticism of intersecting legacies of settler colonialism and slavery. On the one hand, Charmaine Williams and Shirley Chau identified isolationist individualism as a survival strategy in response to the negative stereotyping of ethnic minorities in the multicultural mosaic of Canada (Williams and Chau 2007, 292). On the other hand, the voices of people of color could either be commodified as multicultural to undermine anti-racist struggles towards equity or included in conservative advocacy (292). The extent to which Adar and Eyob-Austin can position themselves with regard to racism and sexism is thus delimited by the particular self-image of

Canada at the moment of narration. Due to the ongoing marginalization of discourses of Black female resistance such as the outstanding writings and performances of Dionne Brand and Cooper herself, I consider it fruitful to read Adar's and Eyob-Austin's self-presentations in terms of womanist thought although this concept did not originate in Canada, but the U.S. But despite its origins outside the current borders of Canada, the writer Alice Walker's concept of womanism addresses the relationships of Black girls to both, their mothers and their white peers who are rather "girlish" than mature and responsible (Collins 2006, 59).

Indeed, as the following discussions of their life writings will reveal, both Adar and Eyob-Austin reflect their experiences and identities in a critical style that testifies to a sociopolitical awareness well beyond their years. In the first case, Adar counters the Islamophobic trope of Muslim girls lacking agency in their choice whether or not to wear the hijab and which profession to pursue, if any at all. In the second case, Eyob-Austin does not shy away from calling out bias against her Blackness, whether it is practiced by teachers, classmates, or internalized. At the same time, their voices could initiate a cultural conversation among Canadian women and girls of color to whom they could provide aesthetic blueprints. While Adar's and Eyob-Austin's narratives represent their observations as girls, they remain in line with the womanist emphasis on mothering beyond biological motherhood and in the sense of female guidance in emancipatory movements, for which Tubman is a telling transnational example following Walker's definition (Williams 2006, 121). Indeed, Walker can be seen to reference Tubman in her definition of womanism when she stages as an example a conversation between a mother and her daughter, who confidently declares, "'Mama, I'm walking to Canada and I'm taking you and a bunch of other slaves with me'" (Walker 1983, xi). While Tubman helped slaves flee from the U.S. to Canada, the narratives of Adar and Eyob-Austin from Canada support contemporary Black U.S.-American discourses for social justice. Yet, the suitability of womanism as a contextual framework for understanding their self-presentations stems from the near-universal aspect of intergenerational dialogue that structures Walker's initial examples as well.

As explained by Cheryl J. Sanders, the classification of a Black girl's resistance against systemic oppression as womanist typically begins when she rejects her own mother's traditional values of modesty and purity (Sanders et al. 2006, 130–31). Importantly, the girl's protest neither results in a finite rupture with her mother and her convictions nor the absolute submission of any one side, but builds a bridge between past, present, and future efforts to improve personal and communal wellbeing through faith, love, and strength. Despite the seeming universality of the theme of intergenerational difference, however, the author of this article wishes to heed Helen (charles)' warning that the label "womanism" should be applied with caution in light of the regional specificity of its linguistic roots in the Black female experience in the rural U.S. South ((charles) 2006, 364). Against this backdrop, I selected the title “Womanist Awakenings in the Multicultural North” for this article, because it cannot be concluded from Adar’s and Eyob-Austin’s life writings alone whether or not they would define themselves as womanists, but they share its focus on self-defined becoming. In terms of life writing, Audre Lorde’s “biomythography” *Zami: A New Spelling of My Name* (1982) lastingly influenced Black female representations of this theme. Lorde was born in Harlem to parents from Grenada and Barbados and raised under racial segregation before discovering that she was a lesbian and pursuing a degree in library science (Lorde 2018, 7, 13, 80, 156–61). She would become the Poet Laureate of New York State and teach creative writing at universities worldwide. As I will show, the theme of a journey towards self-realization structures the life writings of Adar and Eyob-Austin as well although they are narrating while they are still young.

Reckoning with anti-Blackness in the Northern “Canaan”

What role does the autobiographical coming-of-age format have? Madeleine French’s introduction to the anthology *Black Writers Matter* (2017) – which contains a life narrative by each, Adar and Eyob-Austin, provides an answer. On the one hand, the title aligns itself with the slogan #Black Lives Matter (2013-) of a Black women-led social justice movement from the U.S. on behalf of Blacks affected by police brutality. On the other hand, French describes her vision of “a big dinner where everyone is loosely related, some folks know each other, some do not” but “all have made long journeys to get here; sitting inside of someone else’s experience through the written word” (French 2018, xv). She echoes Paul Gilroy’s concept of a Black Atlantic diaspora that comprises all affected by the slave trade worldwide (Gilroy 1993, 15). In this diaspora, cultural products communicate across linguistic and national borders (6, 32). However, regarding women of color, Amal Amireh and Lisa Suhair Majaj have problematized the reception of their autobiographical texts as authoritative sources on their communities (Amireh and Majaj 2000, 16). They derive the label “*testimonio*” from memoirs of Latina activists that gained popularity in academia in the 1980s (10). Often, readings pit the narrative figurations of female selves against their “othered” groups and enforce stereotypes (7). Therefore, attention is due to constructed selves that defy simplistic identifications as either “First” or “Third World” women (14).

Black women similarly represent a "bicultural" self in their autobiographies, according to María Pilar Sánchez Calle who reads their negotiations of identity aspects within journeys towards fulfillment (Sánchez Calle 1996, 161, 162, 167). For example, she connects Lorde's *Zami* to contemporary struggles against racism, sexism, and homophobia that affected her in the Cold War U.S. (164, 165, 167). Reading *Zami* within the U.S. tradition of Black women's autobiographies, Joanne M. Braxton identifies as recurring themes, familial relationships, journeys towards romantic fulfillment, and a concern for one's community and survival (Braxton 2010, 140, 146). In the case of Lorde's paradigmatic post-Civil Rights movement autobiography, her self-identification as a Black lesbian non-separatist feminist followed recollections of family, school, travels, work, and romances (Lorde 2019b, 138).

Yet, the youth of Adar and Eyob-Austin must be taken into consideration with regard to this theme of becoming. On the one hand, Kate Douglas and Anna Poletti outlined how the life writings of increasingly young narrators have come to satisfy the nostalgia of older consumers for their own bygone years of rage and innocence (Douglas and Poletti 2016, 42). On the other hand, the association of youth with innovation permeates late capitalist cultures and can either result in the exploitative fetishization or dismissive neglect of teenage voices (43–44). Obviously, neither approach would do justice to the life narratives of Adar and Eyob-Austin whose represented identities powerfully embody their lived realities in a particular idea of Canada as shaped by multiculturalism and migration in the twenty-first century. In fact, the populations of Canada and the U.S. alike have significantly increased through immigration since World War II while the number of incoming migrants from countries belonging to the so-called Third World has come to surpass that of Europeans (Lipset 1990, 186). Thus, the typical image of a Canadian or U.S. citizen, for that matter, may very well have to shift from that of a middle-aged white settler to a young migrant of color whose journey towards self-realization epitomizes the ongoing process through which their state constitutes its national body.

Adar's and Eyob-Austin's texts perform the modern purpose of life writing to represent "how one has become who he or she is at a given moment in an ongoing process of reflection" (Smith and Watson 2010, 1). Sidonie Smith and Julia Watson traced the Western practice of writing one's life to the Enlightenment ideal of self-improvement through critical introspection (2). Since the eighteenth century, notable figures have published accounts of navigating struggles to inspire their communities to learn from their examples (4, 13). However, the range of life narratives receiving public recognition reflects the values of the societies they emerge in while life narratives intervene into identity discourses as well. Anna De Fina accordingly recognizes identities as produced by interactive storytelling and not as authentic essences (2015, 352, 361, 363). She distinguishes between the moral self which projects values of the epistemic self upon the agentive self that acts in life narratives in line with Smith's and Watson's distinction between the historical author and the textual narrator communicating with readers (359; Smith and Watson 2010, 16). Nevertheless, Cooper's foreword to *Black Writers Matter* describes the narratives as "written from the first-person perspective" and "extremely personal" (Cooper 2018, ix). This category "First-Person" similarly appears on the websites of the Canadian Broadcasting Corporation (CBC) (CBC News 2021). Yet, *Black Writers Matter* and the CBC do not insist on chronological linearity which is associated with traditional autobiographies by Smith and Watson (Smith and Watson 2010, 6, 11). They acknowledge the diversification of life writing in a globalized and digitized age, but French attributes to "Non-linear expression" "a playfulness" in contrast to the "chronological predictability and hindrance" of the Eurowestern tradition (French 2018, xv).

On the one hand, "playfulness" conjures youth, which is both the subject and theme of Adar's and Eyob-Austin's narratives of negotiating their familial legacies with their Canadian contexts (French 2018, xv). On the other hand, since their recorded narratives were published by the CBC and the University of Regina Press, the question emerges if/how their self-representations can confront the institutionalized multiculturalism that Cooper and French criticize (Cooper 2018, xi; French 2018, xvi). Indeed, Walker warns that Black women's subversive domesticity was devalued, including her own mother's creativity by filling "the holes in our walls with sunflowers" when other modes of recording life stories were unavailable (Walker 1974, 240–41, 242). She also traces the founding of a Black Shaker commune of women and children by female ministers to cultural recuperation and self-definition instead of the lesbianism they were accused of in post-Civil War Philadelphia (Walker 1981, 77, 80–82) – a dynamic of ennoblement which characterizes the life narratives discussed in the following as well.

While Walker's notion of womanism is thus deeply interwoven with her Christian faith, it must be noted that Adar is Muslim and Eyob-Austin is Jewish and thus neither of them practices Christianity which holds particular cultural significance in the context of Black history in North America. For example, although Adar was already born into a Muslim family with Somali roots, Muslim faith as a collective practice in North America is strongly associated with the Black nationalism that followed in the wake of the Civil Rights movement in the U.S. to advocate pride in African origins and Black corporeality (Curtis 1994, 53). Yet, relations between Black U.S.-American converts to Islam and Muslim migrants have been complicated by mutual stereotypes of the first group as less intelligent and inauthentic and the second group as too conforming to Western culture and authorities, respectively (64). Against this backdrop, Black girls like

Adar who practice Islam as a legacy of their migrant families could face particular isolation by being firstly, excluded from Black nationalist uses of Islam, secondly, rejected by their migrant communities whether they integrate into mainstream society or not, and thirdly, struggle with the specification "Black" in the label "Black feminism" that emerged from the specific Black U.S.-American context. In contrast, Debra Majeed points out that the term "womanism" "does not require the 'preface' of black" because it is already connected to the uniqueness of Black U.S.-American female experiences (Majeed 2013, 257).

Indeed, Majeed identifies strength in the particularism of the womanist views of women who are Black and Muslim but already living in the U.S. (Majeed 2013, 259, 260). Their experiences may differ from those of Muslim women from majority-Muslim societies because they individually choose their faith and how to practice it in North America (259). Although Adar was introduced to Islam by her Somali parents in Canada, this aspect of Muslim womanism in the U.S. could apply to her as well, since the minority status which is shared by her family and Black U.S.-American Muslims points to their difference from dominantly Christian surroundings in both cases. At the same time, it cannot be taken for granted that Adar as a migrant pursuing integration would necessarily identify a liberating potential in Islam because in the North American context, the religion has openly been embraced by Black people as an alternative to set themselves apart from the white mainstream (264–65). Precisely these paradoxes with regard to Blackness, Islam, and migration call for a concept that recognizes individual women's experiences as sites of knowledge formation. At this point, a reading of Adar's example albeit from Canada benefits from heeding Majeed's insistence that the concept of womanism underlines both, the diversity and authority of the views and voices of Black Muslim women on their situations (262, 263–64).

Fatuma Adar: self-discovery by reclaiming roots

While the "About the Authors" section of *Black Writers Matter* mentions Adar's interest in representing the Black diaspora (French 2019, 191), she begins her narrative "Heavy Scarves" by stating that she lacked "a language to explain what I was doing" in the bathroom during a 6th grade school dance (Adar 2019, 41). On the one hand, Adar's simultaneous anxiety and curiosity are reminiscent of Lorde's recollection of how she anticipated her menstruation to produce "a heavy fullness at the root of me that was exciting and dangerous" (Lorde 2018, 89). On the other hand, Lorde celebrated her menstruation, which is inevitable whereas Adar struggles with the choice whether or not to give herself a "couple moments of bareheaded freedom" by removing her hijab (Adar 2019, 43). As she subverts her Somali parents' norms, she compares unveiling to "a villain revealing their true form" (43), but the result as feeling "bald as a mannequin, a sad plastic attempt to look human" (43). Both self-perceptions are negatively predetermined and imply her need for self-definition, a possibility within Muslim womanism, according to Majeed (2013, 257).

As a Black Muslim womanist, Majeed recognized creative opportunities for self-definition in Islamic scriptures, rituals, and communities that center the principle of choice in Western contexts as opposed to familial tradition or institutional authority (Majeed 2013, 259, 263). This contrast that Adar mentions between "relief" when unveiling "at home" and feeling "like someone is watching" at school leads her to ask "'What are you?'" with regard to "the delicate piece of cloth as it sat heavy in my hands" (Adar 2019, 43). Her paradoxical attribution of subjectivity ("you") to her hijab as she contemplates her agency mirrors her rejection of intersecting stereotypes when she juxtaposes, "I didn't want to be a crusader. I didn't want to be a victim" (43). Such tropes of displacement and oppression risk isolating Adar from both her migrant parents "who at my age had 'actual problems . . .'" and her Canadian friends "distracted by the 'barbaric' nature of our customs" (43). Yet, when Adar summarizes how different groups call her "prettier with or without it" although she saw no difference, she points to corporeal objectification through an external gaze projecting on her "so many adjectives" (43, 44). Considering this containment, Adar's speculation about her unveiled hair mirrors the question-and-response pattern as well as the imagery of immobility and a physical burden in Langston Hughes's poem "Harlem" or "A Dream Deferred" (1951).[1]

> I fooled myself into expecting a Rapunzel-like mane to cascade down as I pulled away my scarf, but instead my hair stood still in its place, unmoved by my decision to set it free. I've been duped, exchanged one burden for another.
>
> I had never met my hair. I didn't know how it behaved in the wild. Would it bounce if I danced to the music? Would it fall in front of my face and block me from seeing what others were saying? During Ramadan a few years later, my mother let me relax my hair. Chunks of it fell off within days. It was the first time I truly saw my hijab as a blessing. (Adar 2019, 44)

Adar expresses both "disappointment and relief" regarding her final choice to exit the bathroom sporting "Jeans, silver top, and scarf" which half subverts, half mirrors familial and mainstream expectations (Adar 2019, 44). Her refusal to

1 Hughes's original poem consists of the following lines: "What happens to a dream deferred?/ Does it dry up/like a raisin in the sun?/Or fester like a sore –/And then run?/Does it stink like rotten meat?/Or crust and sugar over –/like a syrupy sweet?/Maybe it just sags/like a heavy load./ Or does it explode?" (Bigsby 2011, 59).

label her choice as "a step forward or backward" contests both, notions of assimilation as progress and of Islamic traditions as primitive while she likens her hijab to "a crown," visualizing personal sovereignty (44).

Like "Heavy Scarves," Adar's 2017 audio recording for the CBC radio special "Up Close: Black Women Making Canada Better" emphasizes how she chose not to reject her Somali heritage despite her experiences of insecurity as a second-generation migrant (Bero 2018). The narrated overcoming of hesitancy to assume writing due to "struggles that I always thought were the result of being Somali" implies a dichotomy between traditions and individualism that is negotiated by her journey to understand her mother's repeated Somali saying (n. pag.), much like Lorde eventually came to echo her mother's words (Lorde 2019b, 134). While postcolonial identities code-switch to negotiate differences, the practice holds a decolonial potential as well, as in the case of Gloria Anzaldúa who embraced her Chicana heritage by writing despite prejudices while she rejected assimilating into monolinguality and passivity in school (Anzaldúa 2015, 163–64). In contrast, Adar's inclusion of a Somali phrase stages cohesion despite different identities akin to the Canadian multiculturalist project since the 1970s (Thobani 2007, 144). In line with the stream of consciousness in "Heavy Scarves," Adar's recording stages her surprise when she was inspired to co-author a musical of which the following sung lines fade in, "In Mogadishu, the sky was speckled/ . . . /But it's hard to see the stars in Toronto/City lights are the stars so it seemed/And when I first came, I was determined/To be what this place taught me to be/And my parents were disappointed/I let this new world change me" (Bero 2018, n. pag.).

While musicals typically attract privileged consumers, a male voice sings these lines in the style of soul and the words critique the dominance of urban lights, thus performing Homi Bhabha's concept of colonial mimicry, in this case, by appropriating a post-Enlightenment subjectivity to express cultural hybridity (Bhabha 1984, 126, 127). This way, colonized subjects both affirm and subvert Eurowestern expectations as in the example of Bengals in the nineteenth century who responded to British Christianization efforts by on the one hand, eagerly collecting Bibles but on the other hand, repurposing their pages for waste or wrappings (133). In Adar's co-written song, the inclusion of another voice than Adar's and the blend of acoustic modes subvert the Canadian multiculturalist ideal as it is not extended to those back home whose "stars" are eclipsed by the First World with its skyscrapers (Bero 2018). Paradoxically, Adar appeals to individualism but invokes the state when she distinguishes between "you" and "Canada" in the third person, "Canada needs you to be unique and innovative and brave, needs you to be yourself, even when it seems that they don't want it, they need it" (n. pag.). This turn to the national context parallels the coming-of-age process she enacts in the final seconds of her recording with her surprise at having discovered the meaning of her mother's saying as "the quiet revolution" – an image that both mirrors and subverts the universalism of Canadian multiculturalism (n. pag.).

Méshama Eyob-Austin: self-expression through matrilineal empowerment

While Eyob-Austin's bio note in *Black Writers Matter* describes her as an "aspiring singer-midwife" (French 2019, 192), she begins her narrative "Uninterrupted" by stating that she has "yet to find an environment where race wasn't an issue" (Eyob-Austin 2019, 53). She distinguishes how an elementary school classmate compared her looks to "poop," how she was helpless in light of racism, and how Black friends labelled her "the 'whitest'" (53, 54). Her paradoxical description of equality as a "privilege" that "could easily be taken away" invokes a hierarchy based on status rather than just color, following Eve Tuck's (Unangax) and K. Wayne Yang's critique that migrants may join settlers by assimilating into North American white society (Eyob-Austin 2019, 54; Tuck and Yang 2012, 18, 29). Yet, Eyob-Austin's self-description as "born in Montreal to two immigrant parents: an Ethiopian mother and a Jamaican father" emphasizes origins over settlement (Eyob-Austin 2019, 56).

According to Tiffany Lethabo King, Canadian discourses foreground settlers and Indigenous peoples while ignoring anti-Blackness (King 2019, 23). The resulting pressure to assimilate is implied by Eyob-Austin's representation of her teacher's outrage at her friends' discussion about a Black vice principal and the teacher's proposed alternative of a girls' empowerment project (Eyob-Austin 2019, 54, 55). But collective empowerment cannot be compensated by individual opportunities, as Eyob-Austin's mother points out during their meeting with the school's principal,

> My mother informed the principal that I wasn't listening to her, I had clammed up mentally and had removed myself from the situation. She told her that her refusal to acknowledge my Blackness was a refusal to acknowledge me because I was speaking to her of my experiences in relation to my Blackness. Blackness is an essential part of me and it is part of what shapes me. Had I not been Black, I would not have gone through all these experiences that have taught me to identify racism even when others don't think that's what it is. My mom then asked me if I had anything to say. I replied yes. I had a lot to say.
>
> Uninterrupted, I detailed my experience as a Black girl in this school. I told my principal about the stereotypes the teachers had about Black students, the lack of patience and tolerance for us, and the feeling that we were expected to fail. As I spoke I felt my strength coming back, along with a sense of empowerment. When I stopped speaking I realized just how quiet the room had become. (Eyob-Austin 2019, 56)

Eyob-Austin's narrated production of silence contradicts Eurowestern settler notions of filling space to assert presence. Her self-expression during the meeting stands in for her friends whose voices were policed to the point that "We no longer felt we could speak freely between each other; we felt stripped of our right to privacy and felt silenced" (Eyob-Austin 2019, 55). Paradoxically, she rejects stereotypes of absent Black parents but reports how her friends "were ecstatic" about her mother's support for her plan to "'put the principal in her place'" which she rhetorically limits (55).

Her group's desire for dignity despite objectification could answer Walker's call for curious truth-telling about Black communities for which she praises Zora Neale Hurston who promoted folklore and measured Black Harlemite heads, proving that they are interesting and capable (Walker 1975, 101). At the time, this example of Hurston's communal advocacy subverted pseudoscience and countered gender expectations along with Eurowestern cultural supremacy (Walker 1979, 88). Analytically, Eyob-Austin also chronicles her body's "shock mode" and connects land dispossessions of Black and Indigenous people, including her ancestors for whom she prays that their "land is being respected and their experiences are being valued and respected," emphasizing her corporeality, space, and community (Eyob-Austin 2019, 56). By further solidarizing with Missing and Murdered Indigenous Women and Girls, she performs Walker's definition of womanism as identification beyond color,[2] so that, as Eyob-Austin envisions, "Canada can truly be the land of the free" (Walker 1983, xi; Eyob-Austin 2019, 57).

Unlike this closing image, Eyob-Austin's life narrative in the CBC News video "'Life outside of blackness': Montreal family reflects on race, identity" (2016) limits its wellbeing to her three-generational family sharing a Shabbat dinner (MacLellan 2016). Although Eyob-Austin is also shown walking a snowy street in daylight with her voice-over recalling, "no one looks like me in my environment," her statements alternate with responses by her grandmother Malefiya Zeleke, her mother Hirut, and her aunt Eleni Eyob at home (MacLellan 2016). The three generations express a chronology from Zeleke hiding her Jewishness in Ethiopia and Eleni and Hirut Eyob being respectively the "only Black person in an all-white Jewish school" and asked "But you're Black, how could you be Jewish?" (n. pag.), thus paralleling the intersectional exclusion of Black women from anti-racism and feminism in the U.S. (Crenshaw 1989, 140).

2 Walker's definition of womanism includes the following lines: "Traditionally universalist, as in: 'Mama, why are we brown, pink, and yellow, and our cousins are white, beige, and black?' Ans.: 'Well, you know the colored race is just like a flower garden, with every color represented.' Traditionally capable, as in: 'Mama, I'm walking to Canada and I'm taking you and a bunch of other slaves with me.' Reply: 'It wouldn't be the first time'" (Walker 1983, xi).

While Adar jumps into her recollection of approaching a friend to co-write a musical (Bero 2018), Eyob-Austin introduces herself with her full name and as "13 turning 14 in May" after Hirut Eyob summarizes her realization that "Blackness is very much a political identity" and therefore not entirely personal (MacLellan 2016). Eyob-Austin advances the connection of her experiences with Canadian multiculturalism by literally repeating her first racial slur, "Tu ressembles au caca" [You look like poop] in French that she had only circumscribed in "Uninterrupted" (MacLellan 2016; Eyob-Austin 2019, 53). Meanwhile, whether Eyob-Austin is familiar with womanism or not, her hand gestures can be considered to perform its universalism by underlining the connection between suffering affecting her and the broader impact, as she firstly points at the camera when enacting how people will point at her and secondly stages cutting and tossing her dreads "around the house" in the past (Walker 1983, xi; MacLellan 2016). Yet, she subversively prioritizes her family as her source of empowerment when, with a nod and a smile, she repeats her parents' words, "We need to talk" before the camera cuts to her at the head of the table with her mother and aunt lovingly gazing at her (MacLellan 2016). Against the image of her family toasting under photographs of the three generations, Eyob-Austin declares with audible awe that "Blackness is beautiful" despite having just enumerated discriminations, for example, by being followed in the mall because she "live(s) and breathe(s) Blackness" but perpetuates matrilineal legacies of self-definition, as recommended by Lorde (MacLellan 2016, comments added; Lorde 2019b, 139).

Black girlhood in Canada: self-definition as contestation

To summarize, this article discussed life narratives by Adar and Eyob-Austin in terms of their cultural work by representing Black girlhoods under the impact of both, journeys towards self-definition and Canadian multiculturalism. As mentioned earlier, their life writings from *Black Writers Matter* relate to the title of the collection as representative which could serve the goals of womanism and decolonization as well. In both life writings, turning points during girlhood structure the self-representations of the narrative figurations from which they emerge with political awareness: in "Heavy Scarves," Adar wonders whether to wear or not to wear her hijab to a school dance and in "Uninterrupted," Eyob-Austin confronts her principal about racism at her school. Both narrators address communal oppression, for example, through contemporary stereotypes but finally adopt positive attitudes towards their cultural heritages. Analogously, Lorde decided to spell her name as "Audre" instead of the way her parents intended while fantasizing about her mother's Grenada but finally chose for herself "*A Carriacou name for women who work together as friends and lovers*" after her lover Afrekete taught her "roots, new definitions of our women's bodies" (Lorde 2018, 297, 303, italics in original). "Afrekete" is also the name of an African goddess whom Lorde pointed out to the white feminist Mary Daly similarly to how she had to "prove" the existence of her ancestors' island of Carriacou to herself at age twenty-six (Lorde 2018, 13; Lorde 2019a, 58).

In Quebec, Première Pauline Marois denied the agency of hijabi women whom she "othered" when she advertised the ban on "ostensible" religious symbols in institutions as ensuring gender equality, separating the state and the church, and honoring the French legacy (Assemblée nationale du Québec 2010). Adar's explicated choice to wear her hijab contests such definitions of Muslim women as silenced by their cultures. "Heavy Scarves" makes a case against Bill 21 "An Act respecting the laicity of the State" which passed the National Assembly of Québec in June 2019 and prohibits individuals with authority from wearing religious symbols in public functions (Québec Official Publisher 2019, 5, 7, 9, 15–16). Poetics also support politics in "Uninterrupted" when Eyob-Austin criticizes her music teacher's plan for "all the students with 'attitude'" to stereotypically play the same instrument albeit in a separate room from their counterparts, that could caricature Canadian multiculturalism (Eyob-Austin 2019, 53). Pierre Trudeau's Federal Response accorded linguistic and judicial self-determination rights to cultural minorities based on their distinctiveness and separateness (Thobani 2007, 163). Yet, as the represented meeting of Eyob-Austin and her mother with the principal visualizes, white privilege can manifest when the argument of inclusion is (ab)used to legitimate external judgments (Eyob-Austin 2019, 55).

These particularities of the Canadian and Quebecois contexts reveal the limited applicability of postcolonial theory to understand the sociopolitical relevance of Adar's and Eyob-Austin's voices. On the one hand, as described by Chantal Maillé, postcolonial feminism can help expose pretensions of colorblindness as well as denials of white privilege compared to women and men of color in Quebec (Maillé 2007, 92). On the other hand, applications of postcolonial theory to the Quebecois context are complicated by the paradoxical image of Quebec as both, an occupied nation and a participant in Western colonialism with its obfuscation of genocide and slavery (97). As a result of this regional ambiguity, postcolonial theory in Quebec appears to be little known or could suffer appropriation, for example, by separatists like Pierre Vallières who even equated the treatment of the French population in Canada with that of Black people in the U.S. in his controversial work *Nègres blancs d'Amérique* [White N-----s of America] (1968) (95, 99).

However, critical attention to the situation of women from non-Western cultures such as Adar and Eyob-Austin can be considered long overdue in the Quebecois context which has not integrated diverse migrant communities into its self-image (Maillé 2007, 97). As summarized by Maillé, even feminist discourse in Quebec privileges the figure of the white working-class woman to the effect that cultural differences among women living there are only discussed as deficiencies to be overcome through assimilation (105, 107). Against this backdrop, it could seem as no surprise that Adar and Eyob-Austin do not further underline the circumstance that their narrated experiences in Canada occur within the context of Quebec where ambiguous understandings of colonialism and discrimination abound. The very fact that they narrate in English and not in French could point to the comparative predominance of discourses that center Blackness and womanhood in the U.S. compared to Quebec as well as Canada. In this sense, Maillé points out that in the U.S. and Great Britain, feminist movements have been more receptive to anti-racist criticisms because the paradigm of race relations has already compelled theorists to connect issues of race to issues of gender (98). These particularities of the Quebecois context suggest the usefulness of Black women's discourses from the U.S., including in this case womanism, to contextualize the self-representations of Adar and Eyob-Austin from Canada while still taking into account such regional differences as outlined above.

Yet, following Tuck and Yang, the occupied Indigenous peoples and slave ancestors whom Eyob-Austin compares, should not have to consider the future of white settlers in their pursuit of decolonization in North America (Tuck and Yang 2012, 35, 36). Thus, Eyob-Austin's statement, "we may as well just create a new system" does not require any specification as to the "we" just like anti-Blackness becomes apparent from Patrice Khan-Cullors's words while disrupting a Democratic convention in Phoenix, Arizona in 2015, "If I die in police custody, know that they killed me" (Khan-Cullors and Bandele 2017, 232; Eyob-Austin 2019, 57). In *When They Call You a Terrorist: A Black Lives Matter Memoir* (Khan-Cullors and Bandele 2017), this Black Lives Matter co-founder references the police killing of Andrew Loku, an unarmed Sudanese father of five in Toronto and reminds us that youth are the future – a hope embodied by Adar's and Eyob-Austin's turns towards continuity and self-definitions (242, 253).

In this regard, the choices of Adar and Eyob-Austin to represent themselves in the forms of life writing and audiovisual recordings deserve attention. As scholars of hip-hop culture have already noted, less sociocultural prestige is accorded to both musical expression and autobiographical writing than to fictional genres (Douglas and Poletti 2016, 58). Yet, Adar's and Eyob-Austin's examples could challenge frequent emphases on the particular aptness of genres of life writing and hip-hop music to empower young Black men to use their voices (60).

In this sense, Njoki N. Wane underlines common origins of Black women despite their different experiences and refers to oral traditions that transmit unique epistemologies (Wane 2004, 151, 152). While these oral narratives have been transferred across generations, the accumulated collective knowledge continues to support the theories of Black women in the present (151). Therefore, the recordings of Adar's voice as well as the faces and voices of Eyob-Austin and her family do not only reflect contemporary media trends, but could revive oral storytelling traditions from the African diaspora in Canada through the website of the CBC. These spoken self-representations could appeal to internet users who might be cultural outsiders to the same extent as to audiences who share their respective oral traditions. Meanwhile, the CBC's lack of specification of any precise target audiences visualizes the mainstream liberal self-image of Canada as a fundamentally multicultural society devoid of any separatism or distinguishing hierarchies.

Adar's recording and Eyob-Austin's video further address Canadian multiculturalism by providing insights into their trains of thought that their physical voices and/or bodies demarcate. While their words reject assimilation by emphasizing cultural differences, the access to cultural outsiders could glamorize separateness, if not exclusion. In contrast to their life writings, Adar and Eyob-Austin express more optimism to the CBC with regard to their lives in Canada, even while performing personal strength – a rhetoric that Rebecca Wanzo warned of (2009, 34, 180). Albeit in the U.S., Wanzo criticizes the mainstream idolization of Black women's survival stories, for example in the case of Oprah Winfrey, for precluding analyses of their specific oppression (42, 72–73, 76, 78). Yet, Adar's and Eyob-Austin's recordings stage a greater proximity to the historical authors than life writings and thus, corporeal expectations could figure as in Sojourner Truth's abolitionist speech "Ain't I a Woman?," during which she famously pointed to her muscular arms to challenge gender stereotypes (Truth 1851, 510). In the twenty-first century, the stereotype of Black women as "superstrong" can equally preclude them from self-care, support, and public concern in situations of vulnerability (Collins 2000, 159) – which is why Adar's and Eyob-Austin's life narratives of becoming despite adolescent insecurities merit recognition.

Works cited

Adar, Fatuma. 2019. "Heavy Scarves." In *Black Writers Matter*, ed. Whitney French, 41–44. Regina: University of Regina Press.

Amireh, Amal, and Lisa Suhair Majaj, eds. 2000. "Introduction." In *Going Global: The Transnational Reception of Third World Women Writers*, 1–25. New York: Garland Publishing.

Anzaldúa, Gloria. 2015. "Speaking in Tongues: A Letter to Third World Women Writers." In *This Bridge Called My Back: Writings by Radical Women of Color*, ed. Cherríe Moraga and Gloría Anzaldúa, 163–72. 4th ed. Albany, NY: State University of New York Press.

Assemblée nationale du Québec. 2010. "Transcription: Point de presse de Mme Pauline Marois, chef de l'opposition officielle." *Assemblée nationale du Québec* online edition, March 25. http://m.assnat.qc.ca/fr/actualites-salle-presse/conferences-points-presse/ConferencePointPresse-4847.html.

Bero, Tayo. 2018. "Up Close: Black Women Making Canada Better." *CBC Radio* online edition, February 21. https://www.cbc.ca/radio/upclose/fatuma-adar-1.4182943.

Bhabha, Homi. 1984. "Of Mimicry and Man: The Ambivalence of Colonial Discourse." *October* 28 (Spring): 125–33.

Bigsby, C.W.E. 2011. "Hansberry Depicts the Struggle of an Emasculated Male Hero." In *Gender in Lorraine Hansberry's* A Raisin in the Sun, ed. Gary Wiener, 56–64. Detroit: Greenhaven Press.

Braxton, Joanne M. 2010. "Autobiography and African American Women's Literature." In *Cambridge Companion to African American Women's Literature*, ed. Angelyn Mitchell and Danille K. Taylor, 128–49. Cambridge, UK: Cambridge University Press.

CBC News. 2021. "CBC First Person FAQ." *CBC News* online edition, March 10. https://www.cbc.ca/news/first-person-faq-1.5927006.

(charles), Helen. 2006. "The Language of Womanism: Rethinking Difference." In *The Womanist Reader*, ed. Layli Phillips, 361–78. New York and London: Routledge. Orig. pub. 1997.

Clarke, George Elliott. 2005. "Foreword." In *The Hanging of Angélique: The Untold Story of Canadian Slavery and the Burning of Old Montréal*, ed. Afua Cooper, xi–xviii. Toronto: Harper Perennial.

Collins, Patricia Hill. 2000. *Black Feminist Thought: Knowledge, Consciousness, and the Politics of Empowerment*. 2nd ed. New York: Routledge.

Collins, Patricia Hill. 2006. "What's in a Name? Womanism, Black Feminism, and Beyond." In *The Womanist Reader*, ed. Layli Phillips, 57–68. New York: Routledge. Orig. pub. 1996.

Cooper, Afua. 2006. *The Hanging of Angélique: The Untold Story of Canadian Slavery and the Burning of Old Montréal*. Toronto: Harper Perennial.

Cooper, Afua. 2018. "Foreword: Performing Miracles: Black Writers Matter." In *Black Writers Matter*, ed. Whitney French, ix–xi. Regina: University of Regina Press.

Crenshaw, Kimberle. 1989. "Demarginalizing the Intersection of Race and Sex: A Black Feminist Critique of Antidiscrimination Doctrine, Feminist Theory and Antiracist Politics." *University of Chicago Legal Forum* 1, no. 8: 139–67.

Curtis, R.M. Mukhtar. 1994. "Urban Muslims: The Formation of the Dar ul-Islam Movement." In *Muslim Communities in North America*, ed. Yvonne Yazbeck Haddad and Jane Idleman Smith, 51–73. Albany: State University of New York Press.

De Fina, Anna. 2015. "Narrative and Identities." In *The Handbook of Narrative Analysis*, ed. Anna De Fina and Alexandra Georgakopoulou, 351–68. Hoboken, NJ: Wiley Blackwell.

Douglas, Kate, and Anna Poletti. 2016. *Life Narratives and Youth Culture*. London: Palgrave Macmillan.

Eyob-Austin, Méshama Rose. 2019. “Uninterrupted.” In *Black Writers Matter*, ed. Whitney French, 53–58. Regina: University of Regina Press.
French, Whitney, ed. 2018. “Introduction: The Act of Gathering.” In *Black Writers Matter*, xiii–xvi. Regina: University of Regina Press. Orig. pub. 2019.
French, Whitney, ed. 2019. “About the Authors.” In *Black Writers Matter*, xiii–xvi. Regina: University of Regina Press.
Gilroy, Paul. 1993. *The Black Atlantic: Modernity and Double Consciousness*. London and New York: Verso.
Hayes, Diana L. 2012. *Forged in the Fiery Furnace: African American Spirituality*. Maryknoll, NY: Orbis Books.
Henry, Natasha. 2020. “Harriet Tubman.” *The Canadian Encyclopedia* online edition, June 24. https://www.thecanadianencyclopedia.ca/en/article/harriet-tubman.
Henry, Natasha, and Andrew McIntosh. 2022. “Underground Railroad.” *The Canadian Encyclopedia* online edition, March 17. https://www.thecanadianencyclopedia.ca/en/article/underground-railroad.
Khan-Cullors, Patrisse, and Asha Bandele. 2017. *When They Call You a Terrorist: A Black Lives Matter Memoir with a Foreword by Angela Davis*. New York: St. Martin’s Press.
King, Tiffany Lethabo. 2019. *The Black Shoals: Offshore Formations of Black and Native Studies*. London: Duke University Press.
Lawson, Guy. 2015. “Trudeau’s Canada, Again.” *The New York Times Magazine* online edition, December 8. https://www.nytimes.com/2015/12/13/magazine/trudeaus-canada-again.html.
Lipset, Seymour Martin. 1990. *Continental Divide: The Values and Institutions of the United States and Canada*. New York: Routledge.
Lorde, Audre. 2018. *Zami: A New Spelling of My Name: A Biomythography*. London: Penguin Modern Classics. Orig. pub. 1982.
Lorde, Audre. 2019a. “An Open Letter to Mary Daly.” In *Sister Outsider: Essays and Speeches*, 57–62. London: Penguin Modern Classics. Orig. pub. 1984.
Lorde, Audre. 2019b. “Learning from the 60s.” In *Sister Outsider: Essays and Speeches*, 129–39. London: Penguin Modern Classics. Orig. pub. 1984.
MacLellan, Ainslie. 2016. “‘Life Outside of Blackness’: Montreal Family Reflects on Race, Identity.” *CBC News* online edition, March 14. https://www.cbc.ca/news/canada/montreal/real-talk-on-race-montreal-jewish-ethiopian-1.3490044.
Maillé, Chantal. 2007. “Réception de la théorie postcoloniale dans le féminisme québécois.” *Recherches féministes* 20, no. 2: 91–111.
Majeed, Debra. 2013. “Resisting the Veil of Universalism: Muslim Womanist Philosophy as a Lens for Authentic Representations of African American Muslim Women.” In *Authentic Representations of African American Muslim Women*, ed. Ednan Aslan, Marcia Hermansen, and Elif Medeni, 247–66. Frankfurt am Main: Peter Lang Edition.
Maynard, Robyn. 2017. *Policing Black Lives: State Violence in Canada from Slavery to the Present*. Halifax: Fernwood Publishing.
Mbembe, Achille. 2010. “Faut-il provincialiser la France?” *Politique africaine* 119 (October): 159–88.
Québec Official Publisher. 2019. “Bill 21 (2019, chapter 12). An Act respecting the laicity of the State.” *National Assembly of Québec* online edition, June 16: 3–18. https://ccla.org/wp-content/uploads/2021/09/2019-06-16-Bill-21-Passes-and-Becomes-Law.pdf.
Sánchez Calle, María Pilar. 1996. “Audre Lorde’s *Zami* and Black Women’s Autobiography: Tradition and Innovation.” *Bells: Barcelona English Language and Literature Studies* 7: 161–69.

Sanders, Cheryl J., Katie G. Cannon, Emilie M. Townes, M. Shawn Copeland, bell hooks, and Cheryl Townsend Gilkes. 2006. "Roundtable Discussion: Christian Ethics and Theology in Womanist Perspective." In *The Womanist Reader*, ed. Layli Phillips, 126–58. New York: Routledge. Orig. pub. 1989.
Sawallisch, Nele. 2016. "Trudel's Legacies: For a Critical Understanding of Slavery in Quebec." *Zeitschrift für Kanada-Studien* 36: 86–101.
Smith, Sidonie, and Julia Watson. 2010. *Reading Autobiography: A Guide for Interpreting Life Narratives*. 2nd ed. Minneapolis: University of Minnesota Press.
Thobani, Sunera. 2007. *Exalted Subjects: Studies in the Making of Race and Nation in Canada*. Toronto: University of Toronto Press.
Truth, Sojourner. 1851. "Ain't I a Woman?" In *The Norton Anthology of Literature by Women: The Traditions in English*. 3rd ed. Vol. 1, ed. Sandra M. Gilbert and Susan Gubar, 510–11. New York: W.W. Norton & Company, 2007.
Tuck, Eve, and K. Wayne Yang. 2012. "Decolonization Is Not a Metaphor." *Decolonization: Indigeneity, Education & Society* 1, no. 1: 1–40.
Walker, Alice, ed. 1974. "In Search of Our Mothers' Gardens." In *In Search of Our Mothers' Gardens: Womanist Prose*, 231–43. Orlando, Austin, New York, San Diego, Toronto, and London: A Harvest Book Harcourt Inc., 1983.
Walker, Alice, ed. 1975. "Looking for Zora." In *In Search of Our Mothers' Gardens: Womanist Prose*, 93–116. Orlando, Austin, New York, San Diego, Toronto, and London: A Harvest Book Harcourt Inc., 1983.
Walker, Alice, ed. 1979. "Zora Hurston: A Cautionary Tale and a Partisan View." In *In Search of Our Mothers' Gardens: Womanist Prose*, 83–92. Orlando, Austin, New York, San Diego, Toronto, and London: A Harvest Book Harcourt Inc., 1983.
Walker, Alice, ed. 1981. "Gifts of Power: The Writings of Rebecca Jackson." In *In Search of Our Mothers' Gardens: Womanist Prose*, 71–92. Orlando, Austin, New York, San Diego, Toronto, and London: A Harvest Book Harcourt Inc., 1983.
Walker, Alice, ed. 1983. "Womanist." In *In Search of Our Mothers' Gardens: Womanist Prose*, xi–xii. Orlando, Austin, New York, San Diego, Toronto, and London: A Harvest Book Harcourt Inc., 1983.
Wane, Njoki N. 2004. "Black Canadian Feminist Thought: Tensions and Possibilities." *Canadian Woman Studies/Les cahiers de la femme* 23, no. 2: 145–53.
Wanzo, Rebecca. 2009. *The Suffering Will Not Be Televised: African American Women and Sentimental Political Storytelling*. Albany, NY: State University of New York Press.
Williams, Charmaine, and Shirley Chau. 2007. "Notes on Feminism, Racism and Sisterhood." In *Theorizing Empowerment: Canadian Perspectives on Black Feminist Thought*, ed. Notisha Massaquoi and Nyoki Nathani Wane, 285–95. Toronto: Inanna.
Williams, Delores S. 2006. "Womanist Theology: Black Women's Voices." In *The Womanist Reader*, ed. Layli Phillips, 117–25. New York and London: Routledge. Orig. pub. 1986.

Sabrina Völz

Exposing Sexual Abuse among the Amish and Seeking Social Change: Misty Griffin's *Tears of the Silenced*

Amish societies maintain a traditional way of life and separation from mainstream American culture. Generally, they are viewed as gentle, peace-loving, rural folks known for their plain clothing. The darker side of Amish culture – problems associated with sexual violence and abuse – has been largely ignored in scholarship. The first book-length study on Amish sexuality appeared in 2020 and includes a single chapter on sexual abuse (Cates 2020).

As an accessible form of literature, memoir empowers even inexperienced writers to create cracks in the dominant Amish narrative and examine trauma-laden experiences and lesser-known perspectives on sensitive matters. Trauma narratives frequently emerge as a way for individuals to come to terms with and make sense of their traumatic experiences stemming from social, political, or historical injustices. In addition to promoting readerly empathy, trauma narratives may also serve a "testimonial function," in which "the individual, with his/her/their specific personal, emotional, physical, and psychological experience of trauma, becomes a synecdoche of the larger trauma" (Caminero Santangelo and Michael 2020, 240). And as Graham and Lucero write in their introduction to life writing, testimonial literature functions as "a strategic form of rhetoric to promote social change" (2020, 131).

Since 2011, at least eight memoirs written by former Amish break the silence on rape, incest, physical and sexual abuse, and other topics related to women's bodies.[1] After introducing Amish religious and cultural practices that harm victims and survivors and make reporting sexual abuse complex, I will analyze Misty Griffin's *Tears of the Silenced: An Amish True Crime Memoir of Childhood Sexual Abuse, Betrayal, and Ultimate Survival* (2020). Specifically, I will explore how Griffin skillfully adapts her plotting and mode of storytelling to the highly diverse backgrounds of her implied audiences. Finally, I will show that she assigns meaning to her suffering and pain, marked by the shift from invisibility and helplessness to agency and social action. In this essay, Smith and Watson's terms for narrative identity will be used, e.g. the "narrating I" or teller of the autobio-

1 Cf. Furlong (2011, 2014, 2022); Gingerich (2014); Bontrager (2014); Griffin (2020); Hershberger (2021); Byler (2022).

https://doi.org/10.1515/9783111551968-010

graphical narrative (Griffin 2020), the "narrated I" (Misty/Emma) or the object of study, and the "ideological I" or the self that is "historically and culturally situated" (2010, 72–77). Watson and Smith further distinguish between the "'real' or historical 'I,'" situated in a certain time and location "unknown and unknowable by readers" (72), from the "narrated I." I will not refer to this level in the paper.

Amish ways

Since a thorough investigation into Amish culture falls outside the parameters of this study, in this section, I concentrate on selected aspects of Amish culture related to the ensuing discussion on sexual abuse. First, I describe general facts about the Amish, their diversity, and the positive but harmful stereotype of the Amish as non-violent people. Next, I treat gender conceptions as well as the rules governing Amish life, before moving on to their beliefs that influence how this largely communal group handles sexual abuse, beliefs that in many regards clash with mainstream American expectations and practices.

First, the Amish with a population of ca. 373,620 reside in thirty-two U.S. states and four providences in Canada. In 2015, two Amish-affiliated settlements were added in Argentina and Bolivia ("Amish Population Profile, 2022"). However, raw numbers may be misleading since the Amish are one of the fastest-growing religious groups in the U.S. They roughly double in number every twenty years ("Amish Population Profile, 2022") and will likely reach one million by 2050 (Guarino 2012). The increase mainly stems from an 85 percent retention rate and the birth of five or more children per family ("Amish Population Profile, 2022"). Amish groups will accept converts, but they do not actively engage in proselytizing. Rapid growth and Amish expansion into new geographical areas mean that an increasing number of Americans will need to acquire knowledge about Amish culture and religious beliefs to foster intercultural co-existence.

Second, it is difficult to make sweeping generalizations about the Amish. Three major schisms of the Old Order Amish in the United States and the over forty subgroupings allow for rich diversity (Kraybill, Johnson-Weiner, and Nolt 2013, 12). The Old Order Amish is the umbrella term for the most conservative groups. The heterogeneity of each settlement depends on their migration history, amount of "communal versus individual orientation," and other factors, including congregation size and age, geographical location, traits of community leaders, regional economy, and political environment (Nolt and Meyers 2007, 168–71, 178–79). Thus, no one-size -fits-all reaction or solution to sexual abuse can be expected. For sexual abuse prevention to have a lasting effect, approaches need to be adapted to each affiliation or even settlement since the Amish do not have a centralized religious leadership that presides over all Amish groups: "Forty-nine percent of all Amish settlements in North America contain only a single church district" ("Amish Population Profile, 2022").

Third, as descendants of the Anabaptists and later the Mennonites, the Amish believe in adult baptism, simple living, and a literal interpretation of the Bible. As pacifists, they reject military service, armed conflicts, and oath-taking. Due to these tenets of their faith and simple way of living, a largely uncritical view of the Amish as a non-violent group has lodged itself in the American psyche since the 1950s (Boyer 2008, 365). The dominant culture's nostalgic view of the Amish has also been linked to the belief in their "superior values and their supposed ability to avoid the pitfalls of modernity" (370). One sexual abuse survivor and memoirist blames "the so-called academic experts on the Amish" for their role in the perpetuation of the stereotype of the "gentle people" (Bontrager 2014, 10). Griffin debunks this commonly held view.

Fourth, families play a central role in Amish life, and gender roles are mostly hierarchal, complementary, and fixed. God is at the top of the patriarchal hierarchy followed by the religious leaders, men, women, boys, and then girls (Johnson-Weiner 2020, 96). Children are taught to be reserved and quiet and to "'give up' and 'give in' at an early age" (Kraybill, Johnson-Weiner, and Nolt 2013, 101). To impart these lessons, Amish culture allows for corporal punishment and teaches that the child's will must be broken because children who become too engaged in self-expression and individuality will "eventually leave the church or wreak havoc within it" (Kraybill and Bowman 2002, 188). These factors leave children vulnerable when sexual violence occurs.

Next, Amish ways of knowing are highly guarded and slow to change in comparison to the dominant culture. Crucial for preserving their way of life and group identity, the Amish rely on a shared heritage, a common language, and the *Ordnung,* which is defined as a "set of moral guidelines formulated by the local church and adapted over time" (Kraybill, Johnson-Weiner, and Nolt 2013, 42). These rules prescribe everything from Amish hairstyles, dating rituals, and the design of Amish transportation to the use of Pennsylvania Dutch, a Germanic dialect spoken by Plain groups. At the same time, the *Ordnung* in the most conservative groups prohibits numerous practices, ranging from attending high school and filing a lawsuit to using the internet and owning electrical appliances and automobiles. On a whole, Amish groups are reluctant to modify the *Ordnung* but will adapt their way of life after careful consideration when it is viewed as beneficial to the group and readily reconcilable with their religious beliefs without concern for "feared future consequences" (O'Neil 1997, 1137).

Griffin belonged to the Schwarzentruber Amish, an ultra-traditional and highly change-resistant subgrouping of the Old Order Amish. The Schwarzentruber wear the plainest clothing and adhere to the most restrictive use of technology, e.g. they have no indoor toilets or plumbing. Since the Schwarzentruber Amish "maintain a wall of silence toward outsiders" and do not contribute to Amish directories (Hurst and McConnell 2010, 37–38), uncovering sexual abuse is harder than it already is in other conservative Amish groups.

A controversial subject that has arisen in recent years revolves around the question of how widespread sexual abuse in Amish communities is. It is a question that remains unanswered at this time. When sexual abuse is reported and data collected by the authorities, perpetrators and victims are "not distinguishable by racial category" (Harder 2018, 47); in other words, the existing data entries are not separated from other whites according to religious background. Moreover, no quantitative or qualitative studies on sexual abuse exclusively in this insular group exist. Therefore, the observations below mainly rely on senior scholars in the social sciences as well as health practitioners who have studied and/or worked closely with Amish cultures. It is thus essential that eyewitness accounts in the form of memoirs, podcasts, and blogs be included in studies on the Amish to increase knowledge and begin to address research gaps.

When sexual violations happen, most Amish children neither understand nor have the vocabulary to express what they have endured (Cates 2020, 39). Talking about sexuality is considered taboo, so knowledge about both the body and sex is rudimentary and will vary according to a community's traditions (Johnson-Weiner 2020, 40–41). With only an eighth-grade education that neither includes biology nor formal sex education, young people may not realize that there are times when 'saying no' to unwanted touching or sexual advances is warranted (Hershberger 2021, 67).

Although abuse does happen in the home, Amish youth may also be vulnerable to improper advances and sexualized violence when they leave the sphere of their parents' supervision. After their schooling is finished, Amish youth take on more adult responsibility and work outside the home or community, particularly when their labor is no longer needed at home (Kraybill, Johnson-Weiner, and Nolt 2013, 199). For instance, Amish females may contribute to the family's income as live-in caregivers and domestics in other Amish families, especially when a new baby is born or a parent is ill. Griffin was abused by her community's bishop while she was sent to work as his family's housekeeper and nanny. Viewed by the Amish as preparation for motherhood, Amish teens in such positions receive only a meager allowance: in Griffin's case, three dollars a day (Griffin 2020, 119). Consequentially, young Amish women may lack the financial independence to leave their communities should they need to. This limited ability to acquire or access financial resources would most likely be perceived as economic abuse, i.e. economic restriction and dependency, by domestic abuse prevention experts (Sharp-Jeffs 2022).

When it comes to differing views on sexual abuse, perhaps the greatest challenge is the unresolved conflict between Amish perspectives of sexual abuse as a sin and the dominant cultures' view of it as a crime and/or mental illness to be handled by the legal and social systems. The Amish belief that abuse is a church matter is the most significant barrier to reporting abuse to outside officials. The Amish faith and their way of life are closely guarded by their religious leaders, who often place the forgiveness and salvation of the perpetrator over the victim's mental health and well-being: "Women who are victims of violence (or find their children victims of violence) may literally be told that just as Jesus suffered, we [the Amish] suffer as well" (Cates 2014, 88).

After sexual violence has been uncovered, Amish church practices almost certainly cause victims additional trauma. Due to the centrality that the concept of forgiveness plays in Amish beliefs, offenders may confess the same sin multiple times, be socially sanctioned each time, and be forgiven repeatedly without much changing (Cates 2014, 114). This practice does not fare well with victims and does not serve as a proper deterrent to existing or potential perpetrators. Therefore, as Kraybill et al. conclude, the church is not well prepared to deal with addictive behaviors (Kraybill, Johnson-Weiner, and Nolt 2013, 96). Amish leaders may be more concerned with weighing the decision of whether to report the sexual violence to outside authorities than with the short- or long-term effects of the abuse and public confession on the victim (Cates 2020, 91).

Another corresponding factor relates to the way in which most Amish view health care. Amish groups eschew the welfare state, neither expecting nor desiring federal, state, or local aid. Instead, they rely on God, their family, and their community to meet their needs. Although there are exceptions, the Amish mostly do not participate in health insurance programs, and they tend to prefer home remedies to modern medicine (Kraybill, Johnson-Weiner, and Nolt 2013, 336–39). The Amish belief that healing occurs through "the practices of forgiveness and submission to church authority" may delay or prohibit victims from getting the psychological treatment or health care they require (Cates 2014, 96). In their book, *Sexual Abuse within the Church: Assessment, Intervention, and Prevention*, Burkey, Braswell, and Whitehead support the need for counseling either inside or outside the church not only for victims but also for their families as a means to help them cope with the abuse and to reduce the chances of individuals harming themselves and/or leaving the church (Burkey, Braswell, and Whitehead 2022, 93).

Perspectives between mainstream U.S. society and the Amish also differ on the victims' assumed complicity and the proper ways to treat offenders and victims. The American legal system does not recognize consent to sexual acts under the age of sixteen (AgeOfConsent.net, 2022). In contrast, "the degree of responsibility accorded the [Amish] victim is based on numerous factors, including age, maturity, and level of [alleged] consensual participation" (Cates 2020, 114). Griffin tells the story of a young Amish girl who was sexually abused by a convicted sex offender, whom the congregation saw as their duty to take in: "The men in the community believed Laura was to blame, even though she had been a mere fourteen and Larry a grown man in his fifties. 'She had been too friendly,' I would later hear some of the men remark" (Griffin 2020, 120). Victim blaming betrays the trust of victims and shows a lack of empathy for them and the trauma they experience. It further serves as an attempt to cover up the community's negligence for failing to provide adequate supervision of a sex offender and ensure the safety of a minor.

Next, aware of the potential shame and humiliation of public confession and forgiveness rituals that emphasize obedience and submission, Amish victims may hesitate or refrain from "burdening" their families with taboo offenses, especially if the perpetrator is well respected by the community or closely related to a minster (Hurd 2015, 248). Due to Amish kinship beliefs and unequal power distributions, silently enduring the pain and guilt may be preferred to not being potentially believed and losing community standing. Moreover, victims may fear social sanctions if they do not adhere to religious scripts that prescribe the public forgiveness of the offender when called for in a church service with the entire congregation present (Cates 2020, 60), i.e., victims have little to no say in the timing of forgiveness rituals. Due to confessions of abuse during church services, most Amish members are well aware of sexual abuse going on in their midst.

As a rule, Amish parents reject sexual abuse. Yet, an unknown number choose to do nothing to help their loved ones after a sexual assault occurs (cf. Furlong 2011; Hershberger 2021). If caring family members wish to support victims, the church may actively prevent them from seeking help from cultural outsiders, including police officers, social workers, and/or psychologists (Cates 2014, 95). Amish groups additionally prefer to deal internally with problems out of fear that the state may place children in non-Amish foster care; doing so would expose them to the evils of the world or sins that are believed to endanger an Amish person's eternal destiny (116–17).

The Amish practices described above are commensurate with the manipulation of victims through "religious traditions, religious tenets and religious leaders" related in Amy C. Tishelman and Lisa A. Fontes's study of 42,000 victims and survivors of child sexual abuse from twenty different religions, including the Amish. Tishelman and Fontes conclude that religion was "perversely [used] as a tool of corruption, exploitation, shaming, secrecy, and isolation" (2017, 129). As a religion, the Amish are, therefore, by no means alone in their struggle to deal with sexual abuse. As ardent advocates for change, Burkey, Braswell, and Whitehead (2022) provide churches with a wealth of suggestions about how they might address and resolve problems surrounding sexual abuse within religious settings and faith traditions through prevention and intervention strategies. Smaller, insular religions should be included in the efforts to increase awareness about sexual abuse in their churches as well in the ongoing reform efforts. Literature written by survivors of abuse can be part of that important dialogue.

Matters of style and genre

After considering some divergent views between Amish and mainstream society concerning sexual abuse, let us now – as an introduction to memoir – consider matters of style and genre. Let me begin by offering a brief definition of memoir. All forms of autobiographical writing are places of identity construction as the narrating I factors intersectional identities, relationships, and historical or cultural contexts as well as selected episodes of the narrated I's life into a coherent and self-reflective work of non-fiction (Smith and Watson 2010, 38–42). In contrast to autobiography, memoir is a fragmented form of life writing that focuses from the perspective of hindsight on a segment of a writer's life instead of on a whole life. Memoirists employ a number of storytelling tools used by fiction writers but poetic license is generally not one of them. In other words, memoirs are read as "re-presentation of actual humans' experience," not as a creation of their "own lifelike reality" (Couser 2012, 15).

Griffin adapts her style of writing to the implied readers' needs, knowledge, and backgrounds. Stylistically speaking, in *Tears of the Silenced*, the narrating I mostly employs basic narrative and dramatic tools, such as foreshadowing (Griffin 2020, 157, 204), dialogues, and tension, while steering away from other stylistic devices used in fiction, such as a non-linear plotting, symbolism, and inventive language. Since the Amish have little to no training in literary studies, using a more straightforward style of writing increases the readability of Griffin's memoir for that group as it does for general readers.

In *Tears of the Silenced*, the narrating I alternates between summary (telling) and scene (showing) as all memoirists do but favors summary, especially when describing abuse. For G. Thomas Couser, summary concentrates on the I-then (narrating I) instead of the I-now (narrated I), while condensing past events and allowing for a greater degree of reflection than scene. Showing an episode or scene from the past as it unfolds attempts to recreate the experience for the reader (Couser 2012, 69). It also requires "premium . . . writing skills – the ability to describe and to imitate speech" (69). Survivors of abuse may neither have the words to describe horrific events nor may they wish to share all private experiences. Too much drama and description may not only cause authors to relive traumatizing events while composing their self-stories, it may also wound readers, particularly those who have been subjected to sexualized violence and abuse. Due to the "unpredictable and heterogeneous affective responses readers have to narratives of suffering," Smith and Watson voice ethical concerns about "the problems in staging and participating in acts of witnessing to violence and suffering" (2010, 221). Writing about sensitive, traumatic topics is, indeed, a balancing act that each author will approach in their unique way.

Reader responses expressing similar concerns highlighted above contributed to Griffin's decision to extensively revise her self-published 454-page first version of her 2014 memoir, *Tears of the Silenced: A True Crime and an American Tragedy; Severe Child Abuse and Leaving the Amish.* It was subsequentially taken off the market. The 2020 edition saw a reduction of "more graphic content" or potentially triggering passages for victims and survivors of abuse (Griffin 2020, 325). For instance, in one of the descriptions of sexual abuse before Misty leaves her mountaintop home in the chapter "Forgotten by the World," the narrating I makes the narrated I's trauma visible without alienating sensitive readers or driving survivors to relive the sexual violence they experienced:

> When they were not fighting, Mamma would often bury herself in a romance novel while Brian would call me to the loft for a massage. This was often used as pretext to molest me. My legs would shake as I climbed the ladder. I would sometimes push Brian off me, but he would then get angry, and for the rest of the day he would walk around and find things to beat me for. . . . And so, as I climbed the ladder to the loft, I would let my mind travel to a better place. I had learned to drift away to pleasanter places when things around me got to be too much. Most of the time I was a missionary doctor saving AIDS patients in Africa. . . . It was a great fantasy, but not one that was attainable for a girl no one even remembered existed. (Griffin 2020, 40)

Emotionally-charged words, imagery, and other techniques that might be used to create a sensory experience for readers are avoided. Feelings are often implicitly expressed. The words are simple but the sadness of an invisible young girl left alone with her pain and suffering pierces the heart.

To limit graphic and potentially triggering content, Griffin avoids using an ornamental storyteller's voice and the present tense, a narrative tool that, when adopted, attempts to draw the reader deeper into the narrative. In the excerpt above, readers may also notice that the narrating I forgoes the use of contractions except when reporting dialogue, a practice that continues throughout the book. As many Amish and cultural outsiders, including academics, may have a difficult time believing that sexual abuse is a serious problem among the Amish, this soberer and more objective or formal stance lends credibility to the memoir that reads more like an eye-witness account or testimonial literature at times. And since "memory is a subjective form of evidence that cannot be fully verified externally" (Smith and Watson 2010, 7), Griffin includes a few photos at the end of *Tears of the Silenced* as support.

In one of the rare instances in the book in which the narrating I directly addresses the reader unfamiliar with abuse, Griffin explains why the narrated I was unable to escape until much later:

> I am sure it is hard for you, the reader, to believe any girl would return to these people [her parents], you must understand that I was not a normal girl. I was clothed from head to foot in plain clothing; I did not have any knowledge or experience with carrying on a conversation with outside people; I did not understand anything except our daily life on the mountain. (2020, 82)

Additionally, there are many instances in the book that cannot be dealt with here in which the ideological I explains Amish culture (e.g. Amish marriages, the *Ordnung*, baptism, and social sanctions) to general readers, a further indication of Griffin's concern for the needs of multiple audiences.

At this point, I would next like to address the term "true crime" in the title of Griffin's memoir and introduce a few of the genre's conventions. True crime is generally understood as works of fiction or non-fiction in which the author or main character(s) uncover(s) the riveting or shocking facts of an actual criminal case, usually a violent crime, as well as the motives and actions of those involved. True crime narratives use plot twists and embellish parts of the story to heighten the tension (Cecil 2020, 5). By doing so, they may appear journalistic in style but are essentially a form of "infotainment" (5); their authenticity and the degree of truth transported in the work will thus vary.

Unlike a number of true crime stories, *Tears of the Silenced: An Amish True Crime Memoir of Childhood Sexual Abuse, Betrayal, and Ultimate Survival* is narrated from the victims' perspective instead of that of a journalist or investigator who uncovers the facts of the case as they evolve. To some extent, the use of "true crime" with its exaggerations may work against the integrity of memoir as memoir relies on the narrative being as close to human truth as the subjective narrating I's memory permits (Cline and Angier 2010, 14–15). At the same time, however, the merging of true crime and memoir can be seen as a creative genre crossing, similar to the way that life writing not only stands alone as an "independent genre" but has also become an integral part of many other forms of life writing (Graham and Lucero 2020, 121–22).

Nevertheless, the employment of the term "true crime" as a highly popular genre may partially be responsible for increased book sales, thereby drawing greater attention to physical and sexual abuse in Amish communities. In the preface to *Tears of the Silenced*, Griffin reveals her aim to "raise awareness about child abuse and sexual assault," comfort survivors, and facilitate their healing (2020, n. pag). From a marketing perspective, linking the memoir to true crime, therefore, makes perfect sense. And besides, since the postmodern turn, the boundaries between life writing and other literary genres have become increasingly blurred (Couser 2012, 34).

Finally, using the term true crime in the title underscores sexual abuse as a violation of the criminal legal code that demands human justice. With it, Griffin distances herself from the Amish view of sexual abuse as a transgression committed against God's law that often will be remedied without offenders serving time behind bars. By 2017, *Tears of the Silenced* had already sold over 100,000 copies even though it was not professionally edited and riddled with grammar and punctuation errors (Griffin 2020, 332). Readers learn more about the power of life narratives in her 2020 edition. Some years after Griffin left the Amish, a police officer investigating claims of sexual abuse against the same Amish bishop who had also abused Griffin read *Tears of the Silenced* and gave a copy to a social worker who then passed it on to that bishop's children, most of whom decided to press charges against their father after reading it. The memoir also reveals that the bishop was sent to prison for "molesting nearly all of his eleven children" (328). Despite the fact that the full scope of the effects of life writing on its readers is often unknown, this example illustrates the potential power of even unpolished creative writing to further social change and justice. In that regard, *Tears of the Silenced* is indeed a true crime memoir.

Physical, sexual, and emotional abuse in *Tears of the Silenced*

To delve further into literary analysis, it is important to consider the main theme of Griffin's memoir: abuse. Roughly, the first third of her memoir told in chronological order covers the narrated I's childhood plagued by physical, emotional, and sexual abuse by both her mother Mamma and stepfather Brian in a remote mountain area in Washington state (2020, 1–103). Apart from being beaten about three times a day, Griffin is subjected to severe isolation. The following passage sheds further light on the emotional abuse the narrated I endures and the memoir's title: "One of the worst of his [Brian's] rules was that my sister and I were not allowed to talk to each other or to strangers. The only time we were allowed to talk was when we raised our hands and were given permission. We were also not allowed to play with other children" (21). Griffin was not the only victim in her family to be silenced and abused, just as her stepfather was not the only abuser. The narrated I's mother either joined in or observed the physical abuse of her sister, her aunt Fanny who suffered from mental impairment, and later her demented grandmother and wheelchair user. In addition to failing to provide a safe and nurturing environment, Mamma and Brian neglect Griffin's education. The narrating I estimates that she had the equivalent of a second or third-grade education at the age of eighteen (237), a fact that makes her determination to pass the high school equivalency exam and later earn a college degree all the more inspiring.

Griffin was not born Amish. Her dysfunctional family started to live and dress like the Amish when the narrated I was about six years old. Dressing Amish limited questions from outsiders and increased the sales of the agricultural products and dolls the family produced. All in all, their lifestyle only superficially reflected Amish culture. Griffin's parents were neither baptized in the Amish faith nor attended Amish church services regularly.

Feelings of guilt and responsibility for the narrated I's sister, aunt, and grandmother prohibited Misty from leaving the mountaintop (Griffin 2020, 73). When she finally finds the courage to rebel, Mamma and Brian send the narrated I to live with an Amish family at the age of nineteen. Mamma and Brian falsely believe Misty would fail and return home.

The next roughly one-third of the book (Griffin 2020, 104–234) describes the narrated I's life as an Amish girl. After being informally adopted by an Amish family, Misty must choose an appropriate Amish first name and is given the last name of her adopted family. She becomes Emma J. Schrock. Emma can be baptized and become a church member after a period of learning her community's way of life, rules, and language. Due to the community's distrust of outsiders and Brian's defamation of Emma's character, this highly scrutinized process takes over two years. It is a fairly happy period in her life until the narrated I becomes painfully aware of the rumors circulating that she was "after" her Amish father. One night a bitter argument ensues that creates an "irreparable rift" (193). Since unmarried Amish women may not live alone, the narrated I is sent to live with Bishop Peter, his wife Phyllis, and their children the next day.

Almost immediately, the sexual abuse starts and takes many forms. Bishop Peter "inadvertently" touches her breasts and begins to enter her room unannounced under the pretense of giving her the upset baby. After a month, the bishop starts showing her his genitals: "I gasped in horror as I saw that the entire front of Peter's pants was open. He was not wearing any underwear and he was fully erect. I swallowed hard, and Peter looked at me piercingly as I stumbled backward and ran up the stairs" (Griffin 2020, 207). Apart from constant flashing the narrated I when nobody is around, the bishop would also hug the narrated I, thrusting his erected penis against her body. Meanwhile, Emma suspects Peter of sexually abusing his young daughter Katie and slowly poisoning his wife (210).

The narrating I repeatedly makes the point that both the Amish community and clergy in positions of power enable sexual abuse to continue. Already marked as a troublemaker, Emma subsequently knows that she would be blamed if she came forward to report the sexual violations and that after a six-week punishment, Peter's sins "would never be spoken of again" (Griffin 2020, 209–10). Traumatized, Emma is caught in a situation with seemingly no way out and is afraid of going to hell if she leaves the Amish (214). One evening after months of abuse, Peter brutally rapes her and vows to regularly come to her room during the night (212–13). Making threats is an additional form of emotional abuse that terrorizes Emma. During another violent attack, Emma finds the courage to fight back and decides to seek help from a couple living on the bishop's property. Despite the potential consequences and some confusion about what to do, the narrated I makes her way to live with Brian's sister, Aunty Laura, and her husband, Uncle Bill, who know nothing about her abusive childhood.

The positive stereotype of the Amish as a gentle folk not only affects the Amish themselves and the general public but also those involved in law enforcement who allow the Amish much leeway in the resolving of social problems so as not to impinge on their religious freedom. The narrating I relays several instances when police officers reluctantly agree to investigate the narrated I's reports of abuse (Griffin 2020, 137, 143–45, and 219–24) but only exert a minimal effort:

> Well, I really don't know what we can do except go out and question his family . . ., but I even hate to do that to an Amish bishop. . . . It's just that anyone could make the case that you are an Amish girl that is angry about something and has decided to falsely accuse the bishop of her church of a crime for revenge. (222)

The narrated I further argues that if police officers understood the Amish culture better, they would not question an Amish minor's sincerity when going to the police: "'Of course, you haven't had any trouble with them. . . . Do you think they are going to tell you if someone is raped, murdered, or poisoned?'" (222). It follows then that in order for Amish sexual abuse victims to have a fair chance at justice, police officers in the areas in which Amish live must have the appropriate intercultural training.

After unsuccessfully attempting to press charges, it comes as no surprise that the Amish blame Emma for involving outside authorities:

> I was told to drop [the] charges immediately and repent, but suddenly something came over me. I knew they were trying to silence me like so many victims had been before me. . . . I had to break the chain; I would not be another silenced broken woman who had to carry the burden of what happened to her alone, believing it had been her fault. Was being Amish truly the only way to get into heaven? (Griffin 2020, 226)

This passage marks Emma's turning point. Although the fear of going to hell is a strong argument against leaving, the narrated I talks back to those in authority and reevaluates her religious views. The Amish church not only fails to remove an abusive religious leader from power, the narrating I suggests that the church was instrumental in helping the bishop and his family escape to Canada when the police finally started to close in on him (260–61).

Griffin seems to anticipate the eventuality that some readers might attempt to minimize the abuse depicted in *Tears of the Silence* or criticize the memoir's integrity by accusing the narrating I of exaggerating. Griffin exposes not only Emma's abuse and Laura's case mentioned above but also points to the inability of the Amish to deal effectively with additional instances of abuse. The narrating I shares the story of the bishop's wife, Phyllis. She and her sisters between the ages of twelve and fifteen were molested by her father (Griffin 2020, 163). Phyllis's mother reports her husband to the minister every few years, but each time her husband confesses, he is punished and then forgiven. Later in the book, Misty's sister, Samantha, who also joined Amish, informs the narrated I about "two known sexual predators in her church" (298). Thus, *Tears of the Silence* points to a larger, more serious problem in Amish culture.

Instead of ending the narrative with the point at which the narrated I stands on her own two feet, embracing the world and her transformation as other ex-Amish memoirs written as coming-of-age narratives do (Furlong 2011; Gingerich 2014; Bontrager 2014; Furlong 2022), the last thirteen years of Griffin's life are condensed into one chapter, thereby marginally blurring the border between memoir and autobiography. The final segment of *Tears of the Silence* concentrates on the narrated I's transformation from an Amish outcast and victim of abuse to a young woman who maps out her future and comes to terms with her personal identity after a few setbacks, e.g. recurring traumatic nightmares (Griffin 2020, 278–81), as well as her inability to permanently rescue her grandmother from Mamma and Brian (264–78) and her sister from the Amish (286–309). Aunty Laura and Uncle Bill provide Misty with support and a job while she goes back to school. Later at university, she majors in nursing and begins to write her memoir as an act of healing and empowerment.

The final chapter underlines Griffin's agency as a "voice for the voiceless" (Griffin 2020, 325) and shares reader responses, e.g. that of one practicing Amish reader named Sarah. She reports two further cases of abuse, one involving a family of nine who was abused by the father and another treating a father who molested his young son (332–33), demonstrating that sexual abuse also extends to Amish males. The accumulation of examples of sexual abuse throughout the book and again in its final chapter call *all* readers to action:

> If you are in an abusive relationship or situation, I beg you to take strength from my words and find a way out. . . .
>
> I hope teachers, nurses, doctors, and anyone else that may be in a position to recognize abuse will be more inclined to report it and not merely write it off as an accident.
>
> For anyone else out there, if you suspect abuse, please trust your instincts. . . . Do not shrug it off and tell yourself it is not your problem. I often think it would have only taken one person to save me and my sister from our living hell. . . .
>
> To police officers that read this, I hope that if an Amish person makes his or her way to your station, you believe what is being said. As you can see from my account, the price an Amish person pays for going to the police is severe. None of them would fabricate a story as a way to get back at someone. (335–36)

Although the audience can certainly glean their own insights from the text, Griffin places the responsibility for ending abuse specifically on readers, who have become witnesses to her story. Directly naming each group is a persuasive strategy that personalizes the book and makes it more difficult for readers to distance themselves from the knowledge gained from reading *Tears of the Silenced*. The call for change among the Amish church and its clergy, although they are not specially named in the passage above, is implicit. Sometimes, silence speaks louder than words.

According to her homepage, the e-book of her memoir has now been downloaded roughly 500,000 times (Griffin 2022a). This success likely due in part to the rise of memoir and the fascination with true crime in the late twentieth century as well as the public's fascination with anything Amish has opened up new opportunities for the skilled storyteller. In addition to seeking social change through life writing and maintaining a social media presence, Griffin has been the subject of numerous news articles and interviews. Recently, she served as consulting producer of a two-part documentary series on Peacock, *Sins of the Amish*, which explores eyewitness accounts of four women who as members of the Amish and Mennonite communities experienced violent crimes. Griffin is one of the women featured. The first part was aired on May 24, 2022, as part of True Crime Tuesday. Also in May 2022, Griffin initiated a petition to mandate the reporting of abuse in insular societies and prevention-oriented sex education in homeschool settings and in private schools, the vast majority of which are religiously based. As of September 9, 2022, over 17,000 people had signed it (Griffin 2022b).

Conclusion

Sexual abuse is viewed as a religious matter to be handled within the Amish church and is thus seldom reported to outside authorities. The Amish mainly prioritize concern for forgiveness and the offender's salvation over the needs of the victims. Depending on how the male leadership responds to the particular circumstances of the sexual abuse, the "forgive and forget" mantra has left numerous victims vulnerable to further attacks and trauma. As a registered nurse, writer, speaker, and activist, Griffin creates awareness for those issues and fosters the public's empathy for victims and survivors of abuse. The memoirist makes a compelling case that sexual assault is more widespread among the Amish than many community insiders and outsiders are prepared to accept. She further seeks allies who are willing to work toward ending sexual abuse and protecting children from future harm, especially those living in religious environments.

And although Griffin does not openly address academics in her memoir, I ask that we – if we are not already doing it – consider expanding our syllabi to include lesser-known writers, such as Misty Griffin, instead of exclusively studying established writers and their highly polished works of life writing that permit us to teach a particular theory or creative writing technique. In *Tears of the Silence*, Griffin has demonstrated the power that straightforward non-fictional storytelling can have to bring abusers to justice and enact social change in a world that desperately needs it.

Works cited

AgeOfConsent.net. 2022. "Legal Ages of Consent by Country." https://www.ageofconsent.net/world.

"Amish Population Profile, 2022." 2022. *Amish Studies: The Young Center*. Elizabethtown College, Elizabethtown, Pennsylvania. June. https://groups.etown.edu/amishstudies/statistics/amish-population-profile-2022/.

Bontrager, Torah. 2014. *An Amish Girl in Manhattan: Escaping at Age 15, Breaking all the Rules, and Feeling Safe Again*. New York: Know-T Publishing.

Boyer, Paul S. 2008. "Understanding the Amish in Twenty-First Century America." *The Mennonite Quarterly Review* 82, no. 3: 359–76.

Burkey, Chris Rush, Michael C. Braswell, and John T. Whitehead. 2022. *Sexual Abuse within the Church: Assessment, Intervention, and Prevention*. New York: Routledge.

Byler, Mary. 2022. *Reflections of an Amish Misfit*. 2022. Author.

Cates, James A. 2020. *Serpent in the Garden: Amish Sexuality in a Changing World*. Baltimore: John Hopkins University Press.

Cates, James A. 2014. *Serving the Amish: A Cultural Guide for Professionals*. Baltimore: John Hopkins University Press.

Caminero Santangelo, Marta, and Joachim Michael. 2020. "Trauma Literature in the Americas." In *The Routledge Handbook to the Culture and Media of the* Americas, ed. Wilfried Raussert, Giselle Liza Anatol, Sebastian Thies, and Sarah Corona Berkin, 239–51. New York: Routledge.
Cecil, Dawn K. 2020. *Fear, Justice & Modern True Crime*. Boulder: Lynne Rienner Publishers.
Cline, Sally, and Carole Angier. 2010. *The Arvon Book of Life Writing: Writing Biography, Autobiography and Memoir*. London: Methuen Drama.
Couser, G. Thomas. 2012. *Memoir: An Introduction*. Oxford: Oxford University Press.
Furlong, Saloma Miller. 2014. *Bonnet Strings: An Amish Woman's Ties to Two Worlds*. Harrisonburg: Herald Press.
Furlong, Saloma Miller. 2022. *Liberating Lomie: Memoir of an Amish Childhood*. [Broadway, VA]: Memory Pages Press.
Furlong, Saloma Miller. 2011. *Why I Left the Amish: A Memoir*. East Lansing: Michigan State University Press.
Gingerich, Emma. 2014. *Runaway Amish Girl: The Great Escape*. Progressive Rising Phoenix Press.
Graham, Maryemma, and Mercedes Lucero. 2020. "Life Writing." In *The Routledge Handbook to the Culture and Media of the Americas*, ed. Wilfried Raussert, Giselle Liza Anatol, Sebastian Thies, and Sarah Corona Berkin, 121–35. New York: Routledge.
Griffin, Misty. 2022a. *Homepage*. https://www.mistygriffin.com.
Griffin, Misty. 2022b. "#Invisible- Help abused Children in Religious Communities and Institutions." September. *Change.org*. https://www.change.org/p/invisible-help-abused-children-in-religious-communities-and-institutions?recruited_by_id=c5da3e50-bc1c-11e7-a965-7b3bb41fdd48&recruiter=822508969&utm_campaign=petition_dashboard&utm_medium=copylink&utm_source=share_petition.
Griffin, Misty. 2020. *Tears of the Silenced: An Amish True Crime Memoir of Childhood Sexual Abuse, Brutal Betrayal, and Ultimate Survival*. Vergne: Mango Media.
Griffin, Misty. 2014. *Tears of the Silenced: A True Crime and an American Tragedy; Severe Child Abuse and Leaving the Amish*. Misty Griffin [self-published].
Guarino, Mark. 2012. "For Amish, Fastest-Growing Faith Group in U.S., Life is Changing." *The Christian Science Monitor*, November 30. https://www.csmonitor.com/USA/Society/2012/1130/For-Amish-fastest-growing-faith-group-in-US-life-is-changing.
Harder, Jeanette. 2018. "Keeping Amish Children Safe." In *Child Maltreatment in Insular and Isolated Communities*, ed. Christine James-Brown, Julie Collins, Rachel Adams, Debra Schilling Wolfe, John L. Jackson, Jr., and Antonio Garcia, 45–50. Washington, D.C.: CWLA Press.
Hershberger, Lizzie, with Molly Maeve Eagan. 2021. *Behind Blue Curtains: A True Crime Memoir of an Amish Woman's Survival, Escape, and Pursuit of Justice*. New York: Nauset Press.
Hurd, James. 2015. "The Amish *Gemeinschaft* Community: Pro-Woman?" In *Religion and Men's Violence against Women*, ed. Andy Johnson, 239–49. New York, NY: Springer.
Hurst, Charles E., and David L. McConnell. 2010. *An Amish Paradox: Diversity and Change in the World's Largest Amish Community*. Baltimore: John Hopkins Press.
Johnson-Weiner, Karen M. 2020. *The Lives of Amish Women*. Baltimore: John Hopkins University Press.
Kraybill, Donald B., and Carl F. Bowman. 2002. *On the Backroad to Heaven: Old Order Hutterites, Mennonites, Brethern and Amish*. Baltimore: John Hopkins University Press.
Kraybill, Donald B., Karen M. Johnson-Weiner, and Steven M. Nolt. 2013. *The Amish*. Baltimore: John Hopkins University Press.
Nolt, Steven M., and Thomas J. Meyers. 2007. *Plain Diversity. Amish Cultures and Identities*. Baltimore: John Hopkins University Press.

O'Neil, Daniel J. 1997. "Explaining the Amish." *International Journal of Social Economics* 24, no. 10: 1132–39.

Sharp-Jeffs, Nicola. 2022. *Understanding and Responding to Economic Abuse.* Bingly, U.K.: Emerald Publishing.

Sins of the Amish. [2-part True Crime mini-series] May 24, 2022. Peacock.

Smith, Sidonie, and Julia Watson. 2010. *Reading Autobiography: A Guide for Interpreting Life Narratives.* 2nd ed. Minneapolis: University of Minnesota Press.

Tishelman, Amy C., and Lisa A. Fontes. 2017. "Religion in Child Sexual Abuse Forensic Interview." *Child Abuse & Neglect* 63: 120–30.

Mita Banerjee

Biomediality in the Andes: Life Writing and Disability in Violeta Ayala's and Dan Fallshaw's *The Fight*

What is at stake when we think about life writing and the social in the Americas?[1] Arguably, ours is a time when processes of collective action have sometimes been met with a degree of skepticism. Some critics have argued that it is important to pay attention to the elements that subdivide groups and make up their heterogeneity. Scholars such as Rogers Brubaker have cautioned us against "groupism." Brubaker's intervention is twofold. First, he urges researchers not to adopt the language and self-description of the groups themselves, thus losing their critical distance. Second, he draws attention to the processes in which groups are formed, often in a tension between individuality and collective identity:

> Despite a quarter-century of constructivist theorizing in the social sciences and humanities, ethnic groups continue to be conceived as entities and cast as actors. Journalists, policymakers, and researchers routinely frame accounts of ethnic, racial, and national conflict as the struggles of internally homogeneous, externally bounded ethnic groups, races, and nations. In doing so, they unwittingly adopt the language of participants in such struggles, and contribute to the reification of ethnic groups. (Brubaker 2004, dustjacket)

Arguing against the dangers of groupism, Brubaker has coined the phrase of an "ethnicity without groups" (2004, 1), where social differences such as ethnicity can be taken into account without, however, essentializing them and necessarily associating them with collective action. As noted above, Brubaker emphasizes that academic theory needs to distance itself from such collective movements, rather than adopting the language of their participants (3). Such arguments, it could be claimed, may lead us to look at all political movements with a dose of skepticism. Collective action and the performances that accompany them may thus have come to be viewed with skepticism: they may fail to touch us emotionally or affectively, and they may fail to elicit our solidarity to the extent that we see them as both essentialist and obsolete. In this paper, I would like to explore the implications of this reading to some degree, and to wonder whether to speak

1 I am referring here to the title of a workshop organized by Wilfried Raussert and Susana Rocha Teixeira at the University of Bielefeld, in June 2022. For all their critical comments, which were crucial in reshaping this paper, I am grateful both to the organizers and the participants of the workshop.

https://doi.org/10.1515/9783111551968-011

of the obsolescence of collective action per se may itself be a sign of privilege. Moreover, I would like to ask three interrelated questions. First, I want to inquire what forms life writing can take, and how it might be connected to collective action. Second, I am interested in what shapes collective action can assume, and how the "social" is configured specifically in the Americas. Third, I would like to explore the theoretical concepts that we bring to the understanding of social protest.

It is these considerations that I would like to map onto a specific form of social protest: The "Wheelchair Caravan of Integration" (Brégain 2016), in which disability rights activists embarked on a march of 1520 kilometers across the Andes, traveling from Cochabamba to La Paz, the seat of the Bolivian government. My aim in this paper is to read this march through the methodologies of disability studies, life writing research, and InterAmerican Studies, and to explore how each of these methodologies may complement, but also complicate the others.

The image of crossing the Andes in a wheelchair, without doubt, is a powerful one. Read from a Western perspective and from the perspective of Anglo-American disability studies, it is an image that may invoke individual struggle, resilience, and what has been termed "disability pride." One of the questions that I would like to explore in this paper, however, is whether there is a need for contextualizing disability studies, and whether there may in fact be a necessity for a Global South disability studies, as Anita Ghai has suggested (Ghai 2015). As Grech and Soldatic write in their introduction to *Disability in the Global South*, "Disability studies, and its global North focus and tendencies, have not been too receptive to opening its geopolitical scope and remit" (2016, xiv). To what extent, then, may the Wheelchair Caravan of Integration call for a model of disability that is particularly attuned to the specific Bolivian context from which it emerges?

Accompanying the Wheelchair Caravan of Integration: Violeta Ayala's *The Fight*

It is these considerations that I would like to map onto Violeta Ayala's and Dan Fallshaw's award-wining film *The Fight* (2017). Moreover, I will look at *The Fight* through the perspective of life writing research. The strength of the concept of life writing, arguably, is that it looks at all the ways in which "life" can be "written" or recorded. As Marlene Kadar has suggested, "life writing is not a fixed term, and . . . it is in flux as it moves from considerations of genre to considerations of critical practice" (3). Life writing has hence come to encompass blogs, diaries, autobiographies, but also material objects that may be viewed as a form of self-representation. As I suggest in this paper, life writing may thus also be used to describe filmic documents, with the concept of "writing" being conceived in the broadest possible sense as a form of recording life experiences.

One aspect that the present paper wants to inquire into is the intersection between life writing, on the one hand, and disability studies on the other hand. At the core of the field of disability studies, there has been the issue of representation or rather, of the misrepresentation of lives lived with disability. As Lennard Davis has forcefully argued, persons with disability have had to live up to the frameworks or expectations brought to them by the dominant culture (Davis 1995). In contexts such as this one, the concept of life writing may be especially fruitful, since it implies the individual's taking control over the way in which their own life is being represented.

In the film, life writing can be viewed on two levels. First, Ayala chronicles the life narratives of the protestors who participate in the caravan. This notion of life writing, however, may be problematic in that it ascribes agency to Ayala, the director, rather than to the protestors themselves. The second meaning of life writing that I would like to trace here, is related to this problematic. What if the caravan itself were seen as a form of life writing?

Where Ayala's and Fallshaw's film is key for an investigation of collective action in the twenty-first century is that it resists or at least enhances the logic of Brubaker's notion of "groupism" (Brubaker 2004, 3). Ayala's and Fallshaw's filmic narrative emphasizes that collective action and the recognition of individuality can in fact coexist. Where Brubaker cautions that the focus on the collective may run the risk of suppressing the sense of a group's heterogeneity and hence the individuality of its members, Ayala's and Fallshaw's film skillfully weaves personal narratives into the portrayal of collective action. The film, to evoke Brubaker's own terminology, revolves around "disability *with* groups." Moreover, it points to the specific circumstances of what it means to be disabled in Bolivia.

As an article in *The Guardian* describes the Wheelchair Caravan for Integration, "People with disabilities are among the most discriminated against in Bolivia. Fed up with being ignored, a group of them journeyed across the Andes in wheelchairs and on foot to La Paz in an effort to speak with the president, Evo Morales" (*The Guardian* 2017, n. pag.). The caravan itself arose from a social movement, which originated from a call to action by the disability rights organization "Bolivian Confederation of Persons with Disabilities" (COBOPDI) (Álvarez 2019, x). When the caravan first started, it was a movement of only fifteen participants. By the time it reached La Paz, the seat of the Bolivian government, it had grown to 95 protestors.

The slogan that the protestors chanted as they made their way across the Andes in wheelchairs is itself significant here: "Get to La Paz or die." What this slogan implied was that to the protestors, the life they were currently leading was so precarious that they had little to lose. The urgency of precarity, the precarity of living with disability in Bolivia, could not be more pronounced here. It is for this reason that the caravan was called the "Caravan of Integration": The protestors' goal was to ask the government for a pension of 500 bolivas (72 dollars) so they could live in dignity. It is in these instances that Ayala's and Fallshaw's film skillfully interweaves collective action with individual life narratives. In an interview contained in the film, one of the leaders of the caravan, the young Miguel Mamani states that before he joined the caravan, he overheard his sister-in-law convincing his mother that he might better be put to death:

> Miguel Mamani [lying in his sleeping bag by the roadside]: My sister-in-law said she has a friend who's a doctor. And she asked, "Why don't you inject him? Lay him to rest." It's true, I heard them talking. "We should just inject him," they said. My sister told her husband to find the person who could do it. "Because my brother is suffering," she said. They found out everything. They convinced my mum. With those memories, I came here prepared to die. [The camera shows that the protestors have erected a small camp of tents on the main plaza]. (Ayala 2017)

If the protestors' aim is to "get to La Paz or die," Miguel Mamani's narrative powerfully illustrates what this urgency may mean for specific members of the caravan. In moments such as this one, it may be problematic to invoke a universal model of disability studies or disability rights. Read from the perspective of Western disability studies, the image of crossing the Andes in a wheelchair would be seen as an act of heroism, of individual achievement. Seen from the perspective of Critical Disability Studies, there would be a number of problems inherent in this idea of overcoming. First, it may contain an element of the notion of "supercrip" advanced by Rosemarie Garland-Thomson (2002): the idea that a person is worthy of recognition or admiration to the extent that they "overcome" their disability and perform "feats" that they would not be deemed capable of. As Jan Grue notes, "The central feature of the supercrip is success at *overcoming*, at demonstrating ability beyond that which is commonly expected of disabled people. Such ability may be exhibited in many areas, but the most striking qualification for supercriphood, as well as the most problematic, is physical prowess" (Grue 2015, x). This idea of "supercrip" performances has also been seen by critics to verge on "inspiration porn" (Grue 2016). There may be an element of voyeurism inherent in a non-disabled audience looking at these performances, an audience which may derive inspiration for their own lives. The implication here could not be more problematic: If such achievements are possible for persons living with disability or impairment, how much more might a non-disabled individual achieve? As I suggest in my reading of *The Fight*, Ayala's and Fallshaw's film carefully guards itself against the notion of "supercrip."

In the case of the Bolivian Wheelchair Caravan of Integration, however, it is emblematic of the desperation which drives the members of the march. The tremendous effort of crossing the Andes in a wheelchair is less an act of individual heroism than the sign of collective action. I would argue that it is by carefully interweaving collective action with individual accomplishment that the film guards itself against notions of "supercrip" achievements. Moreover, as I will show in more detail below, the focus is not only on the individual's overcoming adverse condition than on the social and economic conditions which make such an overcoming necessary in the first place. The film is hence much more a form of social critique than a narrative focusing only on individual achievement in its own right.

The notion of the almost superhuman strength it takes for the activists to cross the Andes serves as an indictment of a system that drives individuals to such extreme measures. What this implies is the need for a Global South disability studies, where disability can be configured differently from Western models. One aspect that might be investigated in more detail in this context is whether the balance between social critique and individual achievement that is at the core of *The Fight* may also be related to a notion of disability that goes beyond a Western context. What would be considered physical transcendence and individual heroism in a Western model of disability, would need to be recognized as a desperate effort by disability activists to draw attention to the urgency to their claims.

Moreover, one of the central notions of Western disability studies would be the idea of disability without cure and without cause. In this model, it would be considered a form of ablism to expect an individual to always explain the reason for their being disabled. Moreover, the assumption here would be that once the cause for disability has been explained, disability may in fact be cured. It is against such ableist demands that disability studies have stressed the fact that disability does not need to be "explained" to appease a voyeuristic, ableist public. As Jasbir Puar has argued, on the other hand, such a dismissal of the cause of disability would be highly problematic in non-Western contexts (Puar 2017). For instance, if one lives in a system where the circumstances are such that one may run the risk of being injured at any moment, it may be essential to speak of the cause of disability in order to indict the system in which such injuries can occur (Puar 2017). Such ideas may well be applied to the Bolivian context, where gang violence, poverty, and unsafe work conditions have been rampant. As a human rights report about living with disability in Bolivia has noted, "Official discrimination against people with disabilities is still a reality, in the workplace, in schools, in access to the healthcare system and in access to government services. Societal discrimination against people with disabilities has confined many of them to their homes from an early age, thereby limiting their integration into society."[2]

Far from dismissing the reasons for their having become disabled, the activists portrayed in Ayala's and Fallshaw's film thus describe the causes for their disability: their individual life circumstances living in poverty, with inadequate medical infrastructures, and surrounded by gang violence. By going on a wheelchair caravan across the Andes, the protestors can be said to deliberately stage their disability, drawing attention to the tension between a (political) movement and the plight of being unable to "move" in a society that continues to be inaccessible for people with disability.

Both Ayala and other activist media commentators have stressed that in the context of the Wheelchair Caravan of Integration, the term "movement" is deliberately ambiguous. It connotes both a political struggle – the social movement of the disabled community in Bolivia in their call for a pension for persons with disability – and the literal sense of movement, of being able to move or being restricted in one's movement. According to Álvarez and Neal, "[in the narrative of the leader of the caravan], we see him use the term, *la movilización*, which can be roughly translated into 'demonstration' in English. However, we decided to use the direct translation, 'mobilization,' as it underlines the claims to social, political, and spatial mobility activists in the caravan have been making" (Álvarez and Neal 2019, n. pag.)

At the core of the caravan, it could be argued, there is the social model of disability: the idea that people with impairments are being disabled through the political, social, economic, and material infrastructures of Bolivian society.

Beyond precarity

As noted above, Ayala's and Fallshaw's film carefully interweaves individual narratives into the depiction of collective action. It thus heeds Brubaker's critique of "groupism" and the idea that through the sole focus on groups and collective action, individual life narratives may disappear from view. As if in response to such critiques, Ayala employs, both on a narrative and a filmic level, a technique where wide-angle shots and close-ups are used alternately. With her camera, she accompanies the caravan as it makes its way across the Andes; yet, she constantly

2 http://www.yorku.ca/drpi/files/DRPIBoliviaRepEn.pdf.

zooms in on the individuals that make up the group. Moreover, it is through such close-ups that the film resists the sole focus on precarity. One of the pitfalls of emphasizing the precarity of the protestors, it could be argued, is that they might be seen as lacking in agency, as being mere victims of a Bolivian social and economic system that continues to disenfranchise people with disability. By focusing on individual life narratives of the members of the caravan, on the other hand, the film resists the notion of victimization.

It is the concept of life that, in many different ways, is at the core of the caravan. As one activist puts it, the aim of the caravan was to make the otherwise invisible visible: to show the Bolivian public that people with disability are people like everyone else. As Álvarez puts it, "The activists exposed their message to the media, so people could learn to see that people with disabilities in Bolivia are just like everyone else. They laugh, cry, love, sing, dance, and get angry like any person who does not have a disability" (2019, n. pag.). It is this element of the people's sheer and simple humanity, that the film conveys in one of its most memorable scenes, the conversation between Feliza Ali Ramos and Alex Marcelo Vasquez Bracamonte:

> Marcelo: To go out with Feliza, you have to basically, you're forced to see people with disability through her eyes. You see . . . I'm in love with Feliza. [Lying in her sleeping bag in their tent, Feliza is mumbling something. Marcelo is stroking her face.] Don't cry, you've cried enough.
>
> Feliza: Marcelo, my love, I'm so angry. We've started this fight because of our conviction, believing we could change things for everyone. [Marcelo keeps stroking her tears away, staring straight ahead].

A number of aspects are significant in this passage. First, it becomes clear from the interchange that both Feliza and Marcelo believe in collective action. This faith that collective action may be capable of changing the social order, in turn, is at the core of the Wheelchair Caravan of Integration. To speak of the need for a "disability without groups," to paraphrase Brubaker's idea in a different context, may seem cynical here. What the caravan sets out to resist is precisely the forced individuality and isolation of persons living with disability in Bolivia. In their individuality, it could be argued, the members of the caravan were invisible prior to their joining of the caravan. It is through collective action that they now have become visible and that their voices may finally be heard, if only to a degree.

As the leader of the caravan, disability rights activist Carlos Mariaca Álvarez, recalls, "The caravan, which had received very little coverage in the media and was ignored by government authorities, became known as 'The Invisible March'" (Álvarez 2019, n. pag.). In this vein, the members of the caravan resist the ways in which they have been "written out" of Bolivian civic, economic, and political life.

As one of members of the caravan, Rose Mery Guarita, puts it, "We are talking about children who are abandoned by their parents and people living in extreme poverty. . . . With the protests, we have managed to make the general population more sensitive. I hope the new generation sees us as people, not strange creatures" (Watts 2017, n. pag.).

Within the filmic narrative, the conversation between Feliza and Marcelo is also important for another reason. Their conversation is closely bound up with the fact that they are a couple; as lovers, they take care of each other and make sure that the other is as safe as possible, even under the extreme conditions that make up the 35-day march from Cochabamba to La Paz. Crucially, the scene I have referred to above can be related to recent developments in the field of camp studies. As Bidisha Banerjee has noted, there has been a tendency in camp studies (e.g., in the study of refugee camps in places such as Lesbos or Lampedusa), to focus solely on Agamben's notion of "bare life" (Agamben 1998). This reading would highlight the plight in which refugees would find themselves in these camps, being reduced to their biological life and barely having access to shelter, food, warmth, and hygienic conditions.

Even as such discussions are crucial in driving home the horrific circumstances to which migrants may be reduced in refugee camps, Banerjee notes, there may nevertheless be a pitfall in such depictions. What may disappear from view in such discussions, Banerjee emphasizes, are moments of respite, in which migrants create scenes of tenderness, community, and care even in the chaos which has come to surround their lives (Banerjee 2021, 14). It is such agency of transforming camp life into communal life, I would propose, that is at the heart of Feliza's conversation with Marcelo, as it is depicted in Ayala's and Fallshaw's film. Banerjee's discussion of the need in camp studies for emphasizing refugee's agency and community-building may thus also be applied to the Wheelchair Caravan of Integration. In fact, the protestors' installation of their tents in front of the government palace in La Paz, too, can be seen as a makeshift camp. The fact that Ayala's and Fallshaw's camera should zoom in on such moments of care, of tenderness and even humor, can thus be said to highlight the agency of the members of the caravan.

Ayala's and Fallshaw's film *The Fight* is thus in dialogue with recent depictions of refugee camps that similarly try to strike a balance between an indictment of border policing on the one hand and refugees' agency on the other hand. In light of these considerations, it may be significant that one of Ayala's previous films (*Stolen* 2009) should have focused on human rights violations in a refugee camp monitored by the UN. As Bidisha Banerjee suggests in her analysis of Leila Abdelrazaq's *Baddawi*, a graphic novel about the refugee life of Palestinians in Lebanon, it is important to stress not only the refugees' victimization, but also their creating a full life for themselves even in the direst of circumstances. As she observes,

> Arguing against mainstream understandings of camps as "states of exception" and producers of "bare life" (Agamben 1998), I posit that Palestinian refugee camps like Baddawi are deeply ambivalent spaces of affective attachment, fostering a sense of belonging and identity in complex ways. This article therefore connects camp studies with diaspora studies in the context of forced migration to argue that in order to broker a more empathetic approach to refugees we need to reconceptualize notions of refugee camp spatiality. (Banerjee 2021, 14)

Arguably, the conversation between Feliza and her partner Marcelo is indicative of this very possibility: of forging a life even in a camp, even in the makeshift village of tents that the protestors have erected outside the Plaza Murrillo, the seat of the government, in La Paz. These moments of respite, it could be argued, attest to the extraordinary strength of the protestors who have retained their agency against all odds. As Banerjee has emphasized, it is essential to depict moments such as this one in order to highlight the resistance of marginalized communities to forms of victimization and of dehumanization.

Staging disability

In *The Fight*, Ayala chronicles the protestors' actions in all its minute detail. Once they have reached La Paz, the protestors camp outside the government palace on Plaza Murillo for two entire months. As time progresses, their peaceful protest escalates into a stand-off with the police. There is something acutely painful in watching the police, in their riot gear, confront the protestors. One point that the film highlights is that the extreme measures taken by the police against the members of the caravan stand in direct opposition to the self-image that Bolivia's president, Evo Morales, has conveyed of himself in the public: the idea of himself as a "champion of the poor."

It is at this point that we may also consider the different functions that life writing can assume. Miguel Mamani's life narrative, conveyed as a testimony in *The Fight*, is also an indictment of Bolivian president Evo Morales; in fact, the entire caravan can be seen an outcry against Morales's presidency. As Jonathan Watts observes, "[The film] tells of the struggle for disability rights and benefits that has forced the government of this Andean nation to reconsider its policies and damaged President Evo Morales' reputation as a defender of the poor and oppressed" (Watts 2017, n. pag.). What the caravan sets out to defy and to expose as utter hypocrisy, then, is Morales's assuring the public that people with disability live in complete comfort and utter security in Bolivia. As one of the protestors,

Rose Mery Guarita, points out, such rhetoric could not be more cynical. Guarita describes her life prior to joining the caravan in these terms:

> I went through hell. When I remember everything, there's a force bigger than me inside. The lies the government tells about free health, that people with disabilities live in bonanza, that we have a wonderful life. It's not like that at all. That's what gives me strength to fight. If you've suffered as much as I have, how many people suffer every day? (Ayala and Fallshaw 2017)

Not only does he dismiss the protestors' concerns through constant reference of the "benevolence" of his social politics, but Morales goes on to add insult to injury. He proceeds to turn the protestors into a spectacle that may in fact benefit his own regime in the public eye: He simply defines the members of the caravan as terrorists. As the Deputy Minister for Social Affairs, Alfredo Rada, announces, "Why the demand of disabled people to get into Plaza Murillo? This is obviously a political strategy to destabilize the government. What do they want to do? Take the plaza? Take the government palace? What is behind their plan?" (Ayala 2017).

It is at this juncture that life writing, as the film depicts it, takes a different turn. One aspect that may be addressed in this context is the relationship between life writing, on the one hand and, on the other hand, a physical performance staged to represent a different view of lives lived with disability. Life writing research is hence closely akin to performance studies (Claycomb 2004). So far, I have investigated the life narratives that are at the core of Ayala's film. In the section that follows, I would like to draw attention to the ways in which the protestors use their own bodies, taking life "writing" one step further by moving their own bodies center stage.

Up to this point in the protest, the protestors have simply "borne witness" to their lives and the adverse circumstances that they have been forced to live in. Standing before the police barricades in silence, they attest to their plight simply through the act of staring at the police, waiting to be allowed to speak to the president with both immeasurable patience and desperation. As the wait is prolonged, however, the protestors can be seen to escalate their actions: They proceed to *stage* disability by using their own bodies as a canvas onto which they literally inscribe their demands to the Bolivian president. It is in this context that the concept of "biomediality" may be particularly fruitful for illuminating the specifics of disability performance. According to Jens Hauser,

> In times when performance art – which until recently involved mainly human bodies – shifts towards the more general pattern of performativity in art, artists and performers profoundly redefine what actually is considered a body, human or nonhuman, consequently displacing the focus from its mesoscopic actions to its microscopic functions, from physical gestures to physiological processes, and from staged diegetic time to real performative time. (Hauser 2020, 12)

Ayala's and Fallshaw's film enables us to witness the protestors' use of their own bodies as a powerful tool in representing their demands to the Bolivian government. In this performance, as Hauser notes in a different context, they may be able to "redefine what actually is considered a body" (12): What the members of the Wheelchair Caravan for Integration emphasize in their performance is precisely that they have not been integrated into society, that they have been defined by the general public and by government politics literally as no-bodies, as being less than human.

Moreover, through their performance, the members of the caravan draw attention to the "physiological processes" (Hauser 2020, 12) in which their bodies are unable to move or are only able to move differently. It is on account of these processes that people with disability have been marginalized in Bolivian societies (as well as in Western societies), since large sectors of public life continues to be inaccessible. By drawing attention to these physiological differences, the protestors simultaneously call for respect: they cause the audience to pause, to witness, and to bear this difference in movement, rather than looking away and making the protestors "invisible" in their disability. Furthermore, the idea of "real performative time" is key for an understanding of both the protest itself and Ayala's and Fallshaw's depiction of it. The film does not accelerate, and it refuses to fast forward. As Hauser notes, biomediality is characterized by the shift "from staged diegetic time to real performative time" (12). Through the film's insistence on showing the protest in "real performative time," we, too, have to witness the exertion it takes to cross the Andes in a wheelchair, and the nerve-wracking tension of a two-month stand-off with the police in full riot gear.

If indeed the caravan can be read as a form of life writing, then the protestors can be said to escalate their protest through the performance of their own bodies. What the film drives home with such forcefulness is that the protestors seek to make the invisible visible; and that they proceed to highlight the very bodies which have led to their invisibility in the first place. In the first part of this escalation, the protestors proceed to take the government's words literally: If the government treats them like rubbish, they will dress up as such. Having held out at La Paz for more than 30 days, they confront the president as the human garbage that he says they are:

> Marcelo [to a journalist]: This view the government has of us, to treat us like this, like rubbish, or as if we're disposable, we're going to show them for the last two months, they've treated us like this.
>
> Protestors [chanting]: Evo, Evo, Evo, says that we are rubbish.

In the next stage of escalation, an escalation arising out of sheer despair, the protestors employ their own bodies in yet another sense: If Evo Morales wants to see them naked – the same president who has already reduced them to "bare life" in Agamben's sense by refusing to grant them a pension – they will do just that: They will confront him in all the nakedness of their despair:

> Protestor [speaking into a megaphone]: You want to see us naked? [Some protestors have taken off their shirts. One protestor is in his underwear and has written demands on his naked torso. He is crawling towards the police, who are in full riot gear, holding protective shields.]

This performance, it can be argued, can be seen as a form of life writing. Not only do the protestors perform the nakedness, the vulnerability they have been made to feel, they also inscribe their message onto their naked skin. The government, their performance implies, has reduced them to begging, to having to crawl before both the police and the government; so crawl they will. It is this act of crawling that the film conveys both in wide-angle shots and in close-up. In this context, too, the protestors take Morales's words literally: The government's failure to offer financial support to people with disability in Bolivia, the protestors emphasize, has reduced them to crawling; it has forced them to be "beggars" living at the cost of their families, as Manuel Mamani's narrative so painfully illustrates. If indeed the president wants to see them crawl, the protestors argue through their actions, they will proceed to lie at his feet.

As activists have elaborated in their description of the protest, the performance on Plaza Murillo was also meant to represent those who could not be present in the caravan: the children who cannot afford wheelchairs and are thus reduced to crawling around the house. As one activist puts it, "It is quite painful but we do what we are doing for those children who cannot be mobilized, for those children who do not have a voice to be able to shout what is needed, those children who do not have a crutch, a wheelchair that they need" (Alvarez and Neal 2019, n. pag.). It is in this sense that the performance was meant to shock the public into recognition of the protestors' plight. Megan Neal and Carlos Mariaca Álvarez, one of the initiators of the Wheelchair Caravan of Integration have collaboratively created a website designed for student education in cultural anthropology and disability studies. On this website, they have provided both videos of the protestors' performance in La Paz and study materials. In the video they have uploaded, the protestors' performance is underlaid with music; it is no less painful to watch. At one point, as the protestors crawl over concrete, their bodies start bleeding.

If Ayala's and Fallshaw's film does have a climax, then it is that their documentary proceeds to show the unthinkable: the use of force against the protestors. Eventually, the police use pepper spray, chemicals, and water cannons against the protestors:

> Newscaster [reporting from the scene]: This is the drama we're living at kilometre zero [the protestors are shown, desperately trying to rinse their eyes with water] with the disabled trying to get into Plaza Murillo. They're confronting the police who are using chemical gases.

The "bittersweet legacy" of *The Fight*

The caravan of Integration ends, not in defeat but in mourning. After the caravan had held out in front of the Plaza Murillo for months, a car drives into the protestors in the middle of the night; two people are killed. The last scene of the documentary is a final interview with Marcelo:

> Newscast: A group of 12 people with disabilities were protesting at the viaduct. At 1:30 in the morning, a vehicle crashed through the vigil and ran over almost all of the protestors. Five people were injured. We know of two people who were killed.
>
> Marcelo [to the journalist, crying]: The little girl in a coma is . . . . Her name is Diana, and she's eight years old. It was her mother who died.
>
> Reporter: Was the little girl accompanying her mother in this fight?
>
> Marcelo [crying]: That's how it is . . . . Parents can't separate from their children.

As the government threatens the protestors with ten years in prison, the caravan eventually disperses. After the final interview with Marcelo and before the final credits, there is this postscript:

> To stop the protest, the Bolivian government threatened the leaders of the caravan with 10 years in prison.
> Fearing for their lives, Marcelo and Feliza left the protest.
> Miguel went to hospital after suffering kidney failure during the protest.
> Coming off a 10-day hunger strike, Rose Mery took their case to the UN in Geneva.
> The fight continues.
> In loving memory of the people who died during the protests of 2016. (Ayala 2017)

It is both significant and painful to note that the protest does not end in a victory. This is an outcome that Ayala's and Fallshaw's film does not strive to veil, but that it depicts in all its glaring detail.

Even as the filmic narrative seems to end in the protestors' defeat, there has been a "bittersweet legacy" to Ayala's and Fallshaw's documentary film (Watts

2017). It is here that the role of biomediality may loom large: The film won great international acclaim, and it made the plight of people living with disability in Bolivia visible to a world audience. Eventually and with a long time lag, the Bolivian government did give in to the protestors' demands. Even as *The Fight* ends on a note that could not have been more dismal, its legacy ultimately indicated that it had succeeded in making the voices of the protestors heard:

> Ayala said media coverage had helped, including the release of a trailer for *The Fight*, in February.
>
> "That Morales has met some of the demands of the protesters is a big development for the film," he said. "A year ago, the government was saying that people with disabilities didn't need a pension, while today they are forced to accept the necessity of the pension, which shows the power of filmmaking combined with activism."
>
> The political impact of the protest and the film was stressed by Guillermo Mariaca, a professor at the Higher University of San Andrés.
>
> "The making of the film was very important. It revealed the political and social callousness of the government. And it gave the protesters visibility and political self-esteem," he said.
>
> Mariaca said Morales had been damaged by the scenes of repression, which contrasted starkly with his image as a socialist defender of the poor and indigenous people. "Evo's socialist discourse is lost. Society does not believe that it is authentic any more," he claimed. (Watts 2017, n. pag.)

In its global circulation, Ayala's and Fallshaw's film may thus have achieved what the caravan's local performance could not: the granting of 500 bolivars to people living with disability in Bolivia. At the same time, the "bittersweet legacy" of *The Fight* points to potential of fusing different forms of life writing. As I have tried to show in this paper, the film is a document of life writing because it contains the interviews with and hence the life narratives of the members of the Wheelchair Caravan of Integration. At the same time, it also focuses on the performance by members outside of the presidential palace. The life writing and the activist performance that were inherent in the Wheelchair Caravan of Integration ultimately became globally visible through Ayala's and Fallshaw's documentary film.

Finally, we may want to return to the idea of life writing as a narrative. Would the attempt to read the Bolivian Wheelchair Caravan for Integration as a form of life writing be a "flattening out" of this protest? Would the concept of "performance activism" (Wiegmink 2011) be more apt in describing the Wheelchair Caravan of Integration? I would like to suggest that these concepts may in fact be mutually complementary. Ultimately, this is where an interdisciplinary perspective may be crucial. The Wheelchair Caravan of Integration can be located at the intersection of disability studies, hemispheric American studies, Latin American studies, cultural anthropology, performance studies, and life writing research. Yet, the lenses we bring to reading and understanding the caravan and the social movement it is embedded in, are only valuable as long as they succeed in conveying what may perhaps be central to academic research: the rootedness of scholarship in lived experience.

What makes Ayala's and Fallshaw's film so powerful, it could be argued, is that it frames the protestors' lives and through such framing, makes them visible. At the same time, however, I would like to suggest in closing that the notion of life writing may fall short of its full potential if we attribute the notion of authorship solely to Ayala as the director of the documentary film *The Fight*. As I have tried to suggest in this paper, the protestors' action – as a form of "activist performance" (Wiegmink 2011) – and their use of their own bodies can itself be seen as a form of life writing. In their performance, the protestors draw attention to what their bodies cannot perform; without a wheelchair, they are literally left to having to crawl. With acts such as this one, in the moment of getting out of their wheelchair and crawling across the concrete in front of the presidential palace, they draw attention to their demands: Without the proper infrastructures, without the material means to buy a wheelchair or have access to material infrastructures that make a decent life possible, their lives become unlivable. The performance becomes a form of life writing because it "writes" or inscribes, through physical action, how the members of the caravan perceive their own lives and their dependence on economic subsidies that they are now calling for in front of the president's palace.

At the same time, however, it may be important to go beyond the film itself. Through what other media and in what other channels would the lives of the protestors be visible? At this juncture, it would be important to expand the notion of life writing from documentary film to social media. As Marlene Kadar argued, life writing can take many forms and shapes, it can be written, filmed or practiced (Kadar 1992; Hornung 2010). Recent studies have shown that social media can be crucial in enhancing our notion of life writing (Özkan 2021). As Aimée Morrison observes,

> Everyday subjects perform autobiographical acts online as a condition of social media participation. Such autobiographical acts can be understood to constitute micro-acts of life writing that accrete over time into forms of memoir, or into cross-sectional purposive political speech through hashtags, for example, or which articulate new forms of social identity and social belonging through the development of rhetorical genres attentive to unmet expressive needs. (Morrison 2019, 41)

It is through social media that we may learn about the lives of Feliza Ali Ramos and Alex Marcelo Vasquez Bracamonte in their own voices.

Moreover, it is also important to note that in a newspaper article that she co-authored with Bolivian journalist Andrea Monasterios, Ayala herself talks about the lives of Feliza and Marcelo Vasquez Bracamonte in more detail. Feliza Ali

Ramos is a social worker who was on her way to Uyuni when she was caught in a bus accident (Ayala and Monasterios 2017). She has pointed to medical malpractice as the reason for her disability:

> When she arrived at the hospital, the doctors needed the signature of a family member to operate, but her parents were a three-day bus ride away. Despite the fact that they wouldn't operate, the doctors x-rayed Feliza against her will, she didn't want to be moved, she knew her opportunity to walk again depended on not being moved. *"It was medical negligence, I felt my vertebra pushing against my brain."* When the nurses moved her body into position on the x-ray table, Feliza felt her spine give way with excruciating pain, the worst pain she'd felt in her entire life. She knew she'd never walk again. (Ayala and Monasterios 2017, n. pag.)

In this context, it would be highly problematic to argue that the cause for a person's disability should not be mentioned. In Feliza's narrative, a number of aspects point to the necessity of contextualizing disability and to call for a disability studies of the Global South. In a disability studies focused primarily on the Global North, the focus has often been on the individual and their interaction with a society that continues to be ableist. Moreover, the notion of a "disability without cause" has been central; it has been argued that to ask why a person is now living with disability or impairment is itself part of an ableist regime in which individuals with disability are constantly asked to "explain themselves" (Grue 2015, 204). As important and beneficial as this may be in the context of the Global North, we might argue that the notion of the causes for disability may be differently conceived in the Global South. As *The Fight* illustrates, some members of the caravan have become disabled due to adverse economic and political conditions; Miguel Mamani, as the film describes, has become disabled because he accidentally got hit by a bullet in a gang fight. In contexts such as this one, it would be highly problematic to speak of "disability without cause." Rather, it is important to highlight the cause for a person's disability in order to draw attention to social, political and economic circumstances that need to be redressed.

As the film makes clear, this is as true of Miguel Mamani's life narrative as that of Feliza Ali Ramos. Feliza's becoming disabled is itself tied to specific circumstances in terms of infrastructure, public transport, and adequate health care. As Ayala and Monasterios describe in an article that could also be read as a biography of Feliza Ali Ramos, it was on the day of her accident that she proceeded to rename herself:

> On September 23rd of 1995 Lili died and was born as Feliza, her real name. She left behind an identity that wouldn't belong to her any longer and was forced to start a new life. Days before, a clairvoyant read her hands and closed them without wanting to tell her what destiny had planned for her. *"What did you see? Am I going to die? Is it an accident?"* She asked

> half jokingly but half serious to the clairvoyant. Little did she know that she was guessing her future. (Ayala and Monasterios 2017, n. pag.)

This act of renaming herself, she has noted in retrospect, sparked off Ramos's life as a disability activist. It was at a disability rally that she met Marcelo Bracamonte, a lawyer, who had contracted polio as a child. They got married in 2014 (Ayala and Monasterios 2017).

These aspects may seem significant for a number of reasons. First, they point to the need to expand the notion of life writing beyond published autobiographies or documentary film, and to take into account the role of social media in portraying life narratives. Such a perspective on the lives – and the life writing – by Feliza Ali Ramos and Alex Marcelo Vasquez Bracamonte would be important also because it would restore the notion of authorship to the members of the march, rather than ascribing authorship only to the documentary filmmaker Violeta Ayala, as powerful as her film may be. Moreover, the focus on the lives of Feliza Ali Ramos and Alex Marcelo Vasquez Bracamonte would also shed a different light on the notion of precarity: Both Ramos and Bracamonte are highly educated, and they have been at the forefront of collective action by disability groups in Bolivia. "Precarity" is hence itself a complicated, multi-layered term, a notion through which we may want to re-read Ayala's and Fallshaw's film. As people living with disability in Bolivia, Ramos and Bracamonte may be marginalized, but their middle-class status and their education may enable them to speak for other members of the community, who may be unable to make their needs and demands heard. Finally, the power of Ramos's and Bracamonte's life writing narratives, both on social media and as they are contained in Violeta Ayala's and Dan Fallshaw's documentary film *The Fight*, is that they circulate transnationally. It is through this transnational circulation that they connect the specificity of social protest in the Americas with other forms of social protest worldwide. In debates surrounding the release of the film, for instance, much attention was paid to the fact that "elections in neighbouring Ecuador brought to power the world's first paraplegic president, Lenin Moreno, who has been a vocal campaigner at home and abroad" (*The Guardian* 2017). Such discussions may cause us to think about the way in which disability is conceived and framed in different national and cultural contexts. In this circulation, it may cause us to rethink what we mean when we speak of disability, of precarity, and of collective action. It may also caution us against believing that any of these terms are universal. It is for this reason that the Wheelchair Caravan of Integration is such a powerful call for collective action against social injustice, in Bolivia, in the Americas, and much beyond.

Works cited

Abdelrazaq, Leila. 2015. *Baddawi*. Charlottesville: Just World Books.

Agamben, Giorgio. 1998. *Homo Sacer: Sovereign Power and Bare Life*. Palo Alto: Stanford University Press.

Álvarez, Carlos Mariaca, and Megan Neal. 2019. "Visualizing Disability: Wheelchair Caravan Across Bolivia." *Teaching and Learning Anthropology*. https://teachinglearninganthro.org/visualizing-disability-wheelchair-caravan-across-bolivia/.

Álvarez, Carlos Mariaca. 2019. "Wheelchair Caravan in Bolivia." https://teachinglearninganthro.org/wp-content/uploads/2021/03/2011-Bolivian-Caravan-Printout-.pdf.

Ayala, Violeta, and Dan Fallshaw, dir. 2017. *The Fight*. Documentary.

Ayala, Violeta, dir. 2009. *Stolen*. Documentary.

Ayala, Violeta, and Andrea Monasterios. 2017. "My Right to Live with Dignity." *Huffington Post*, December.

Banerjee, Bidisha. 2021. "Picturing Precarity: Diasporic Belonging and Camp Life in Leila Abdelrazaq's *Baddawi*." *Journal of Postcolonial Writing* 57, no. 1: 13–30.

Brégain, Gildas. 2016. "Comparative Study of Two Protest Marches for Disabled People's Rights (Spain 1933–Bolivia 2011)." *Moving the Social. Journal of Social History and the History of Social Movements* 56: 115–40.

Brubaker, Rogers. 2004. *Ethnicity without Groups*. Cambridge: Harvard University Press.

Buell, Susan. 2013. "A Perspective from Bolivia on the Implications of the World Report on Disability for People with Communication Disabilities." *International Journal of Speech-Language Pathology* 15, no. 1: 32–36.

Claycomb, Ryan Matthew. 2004. *Playing at Lives: Life Writing and Contemporary Feminist Drama*. College Park: University of Maryland.

Davis, Lennard. 1995. *Enforcing Normalcy: Disability, Deafness, and the Body*. New York: Verso.

Garland-Thomson, Rosemarie. 2002. "The Politics of Staring: Visual Rhetorics of Disability in Popular Photography." *Disability Studies: Enabling the Humanities*, ed. Sharon Snyder et al., 56–75. New York: MLA.

Ghai, Anita. 2015. *Rethinking Disability in India*. New Delhi: Routledge.

Grech, Shaun, and Karen Soldatic, eds. 2016. *Disability in the Global South*. Freiburg: Springer.

Grue, Jan. 2016. "The problem with inspiration porn: a tentative definition and a provisional critique, Disability & Society 31, no.6: 838–49, DOI: 10.1080/09687599.2016.1205473.

Grue, Jan. 2015. "The Problem of the Supercrip: Representation and Misrepresentation of Disability." *Disability Research Today*, 204–19. New York: Routledge.

Hauser, Jens. 2020. "Microperformativity and Biomediality." *Performance Research* 25: 12–24.

Hornung, Alfred, ed. 2010. *Autobiography and Mediation*. Heidelberg: Winter.

Kadar, Marlene, ed. 1992. *Essays in Life Writing: From Genre to Critical Practice*. Toronto: University of Toronto Press.

Morrison, Aimée. 2019. "Social, Media, Life Writing: Online Lives at Scale, Up Close and in Context." In *Research Methodologies for Auto/biography Studies*, ed. Kate Douglas and Ashley Barnwell, 41–48. New York: Routledge.

Özkan, Damla. 2021. “Mental Illness Life Writing on Social Media: The Role of Social Media in the Re-Evaluation of Identity for Individuals Diagnosed with Depression.” Alternate, Virtual, and Augmented Realities: A Postgraduate & Early Career Research Conference. 20 September, Durham University. Online Conference Presentation.

Puar, Jasbir. 2017. *The Right to Maim: Debility, Capacity, Disability*. Durham: Duke University Press.

The Guardian. 2017. “The Fight: Disability Rights Protestors in Bolivia on the Barricades.” May 5. https://www.theguardian.com/news/2017/may/05/the-fight-disability-rights-protestors-in-bolivia-on-the-barricades.

Watts, Jonathan. 2017. “Bolivia’s Caravan of Courage Leaves a Bittersweet Legacy for Disabled Protesters.” *The Guardian* online edition, May 5. https://www.theguardian.com/global-development/2017/may/05/bolivia-caravan-of-courage-bittersweet-legacy-disabled-protesters-the-fight-documentary.

Wiegmink, Pia. 2011. Protest EnACTed: Activist Performance in the Contemporary United States. Heidelberg: Winter.

Anna Ferrari

Dancing at the Border of Fiction and Fact: David B. Feinberg's Life on the Page between Activism and Literature

Introduction

After the AIDS epidemic exploded in the United States in 1981, the fact that the disease was sexually transmitted and that – at least at first – it mostly hit gay men sparked a reaction of repulsion and homophobia in the general public. From the beginning, when it was known as "gay cancer," AIDS became synonymous with stigma, resulting in further oppression for the gay community. Within the community, however, the relationship with the virus was more nuanced: PWAs (people with AIDS) were not as subject to marginalization – however, particularly in the 1980s, the epidemic brought a seismic shift in gay people's relationship with sexuality. Literature became an important tool to respond to the challenges of the new reality. While authors embraced different genres – novels, plays, short stories, essays, memoirs – the autobiographical element has always been central to AIDS writing. As institutions, media, and public attention alike surrounded the epidemic with silence because of the stigma attached to it, gay authors strived to portray the human experience of PWAs, be it in form of testimony or protest. In this literary landscape, a particularly interesting approach was chosen by New York writer David B. Feinberg, author of two autobiographical novels between the late 1980s and the early 1990s, *Eighty-Sixed* (1989) and *Spontaneous Combustion* (1991), and a posthumously published non-fiction collection, *Queer and Loathing* (1994). The presence of the virus is pervasive in his production and represents the central theme in his work. Unlike older AIDS authors such as Larry Kramer or Edmund White, Feinberg started to write after the explosion of the epidemic and was immersed in its context for the entirety of his artistic life. The defining trait of his style became, despite the bleakness of the topic, a combination of humor, fury, and pain. The way he chose to deal with the epidemic put him in a unique place in the landscape of AIDS literature, producing intense, highly entertaining pieces of writing. His approach to AIDS is at the center of this analysis.

As aforementioned, autobiographical influences represent a defining trait of AIDS writing. Authors were portraying personal experiences – either as PWAs, activists, caregivers or, more often than not, all of the above – which would have otherwise been invisible to the general population. The writers' involvement in the events described – and therefore the fact that their own survival was on the line –

https://doi.org/10.1515/9783111551968-012

made these texts profoundly personal. Memoirs and essays were employed by many authors, as the filter of fiction to some had seemed inappropriate to tackle the tragedy of AIDS.[1] David Feinberg is not among them, as most of his work – two of his three books – consists of novels. His interpretation of fiction, however, borders on non-fiction: the main character in Feinberg's novels is his alter ego, B.J. Rosenthal, who goes through many events that we know actually took place in the author's life. B.J. therefore functions as quite a thin filter between fact and fiction, between the author's writing and his own life. In Feinberg's life writing, to use Hermione Lee's definition, the distinction between fiction and autobiography is "deliberately blurred" (Lee 2005, 100). As we will see, the distinction between fact and fiction does not become clearer when Feinberg moves to non-fiction as, besides the narrator's name, the author's voice does not show a change from his previous first-person narratives. In both cases, the feistiness of his voice sparks as a reaction to the AIDS epidemic. I will analyze Feinberg's work focusing on how he tried to protect himself from AIDS by putting filters between himself and the virus, from his use of an alter ego to the style that pervades his writing – that is, humor. Especially in his two works of fiction, Feinberg allows us to reflect on AIDS representation in a new way as we follow the adventures of B.J. Rosenthal through an AIDS-infested New York while the author carries us through the trauma of the epidemic with a "wickedly funny" tone, as Catherine Texier called it in her review of *Eighty-Sixed* in *The New York Times* (Texier 1989). We'll see how Feinberg tries to deal with AIDS (going through the stages of fear, shock, and rage) by using humor to protect himself from dealing at face value with the horrors of sickness and stigma. His humor is fueled in equal parts by fear of death and rage against a government that had allowed the epidemic to get as bad as it did. By analyzing the function of humor in Feinberg's work we will see how, despite his attempt to use it as a tool for self-protection, humor ultimately allows him to portray the reality of AIDS in a much more explicit, crude way compared to most AIDS authors, as the filter he tried to put between himself and the virus ultimately reveals itself to be, rather than a protection, a magnifying glass. Humor allows Feinberg to discuss even the most uncomfortable aspects of life as a gay man during what he calls "the Age of Anxiety."

1 Several writers could not bring themselves to write AIDS fiction because, to gay men, AIDS "is a gravely personal issue. It is too real to be easily metaphorized or elegantly aestheticized" (Nelson 1992, 1).

Filters

Considering how personal the involvement in AIDS writing was for gay authors, one's personality was not irrelevant to the shaping of a literary response. Feinberg, by all accounts of those who knew him, was feisty, uncompromising, and restless. These traits are featured prominently in his work, and have become a trademark of his style. The two novels, narrated in the first person, established from the beginning a close relationship between author and character, so much so that the two became impossible to separate: they live in the same neighborhood, have the same mannerisms. Consequently, in Feinberg's case the autobiographical influence has become a particularly prominent element in his work. G. Thomas Couser notes how "the line between fiction and nonfiction, novel and memoir, is sometimes fuzzy, and it is sometimes purposely crossed" (Couser 2021, 13). David Feinberg's work lives on that line. It is worth noting, though, how despite the raw honesty of his writing he only managed to deal with AIDS through filters.

Feinberg's tormented nature brought for him the need to portray the epidemic in a raw, explicit, non-romanticized way. In order to do this, however, he also decided to observe these experiences through a lens, something that would work as a defense mechanism against the traumatic times he lived in. Humor is employed at the same time as a shield to protect the author's sanity and as a weapon against everybody else. Although they have been published in the short span of five years, each of his books represents a different, significant step in his life during the epidemic. The first novel, *Eighty-Sixed*, remains his most acclaimed work and portrays two different eras: the first part is set in 1980, a time before the epidemic and still marked by the sexual liberation, and the second part in 1986, when AIDS had already been raging for a few years. Here, we find B.J. caring for Bob, a former trick who is dying, and obsessing over whether he has been infected himself. In the second novel, *Spontaneous Combustion*, B.J. deals with his own diagnosis. In the final book, *Queer and Loathing*, Feinberg combines the rage about the epidemic in general with the fury about the fact that *he* is dying. In the following pages we will see more about the use of the alter ego as a filter and how it got dropped in the shift to non-fiction. Here we will acknowledge the aforementioned presence of another filter, recognizable throughout all of his work, something used at the same time to keep the trauma of AIDS at a distance and to observe it at face value in order to be able to deal with it: the filter of humor. Although Feinberg is not the only author to employ humor to deal with AIDS, *Eighty-Sixed* is among the texts which introduced more forcefully the idea of laughing about the epidemic. In her review, Texier notes how Feinberg's prose "stands out for its frankness, ferocious wit and total lack of sentimentality or self-pity" (Texier 1989). As we will see, apart from sheltering him from trauma, the employment of this specific kind of humor carries specific implications.

Humor: filter or lens?

As aforementioned, the use of politically incorrect humor has a crucial role in Feinberg's work, as it is the tool he uses as a filter to protect himself from the trauma of the epidemic. The theme is so central in his writing that he dedicates an essay to it, contained in *Queer and Loathing*. The essay, titled "AIDS and Humor," is arguably the closest Feinberg ever came to writing a manifesto: "humor is a survival tactic, a defense mechanism, a way of lessening the horror. I would probably literally go mad if I tried to deal with AIDS at face value, without the filter of humor. Once you joke about something, you appropriate it, you attain a certain amount of control over it" (Feinberg 1994, 87). The act of regaining control, however, did not have for Feinberg the mere effect of self-protection against the trauma of the epidemic. By using humor, Feinberg also portrays the reality of AIDS in a particularly merciless, raw way. In this sense, humor functions less as a lens and more as a magnifying glass, pointing out the suffering and the incongruity of living life as a gay man in 1980s New York: "there is no way to tell us apart from the general population because *we are the general population*. It's Night of the Living Dead, with pod people everywhere" (Feinberg 1994, 26). In his introduction to *Queer and Loathing*, Tony Kushner notes how "Feinberg makes you laugh from precisely the place inside of you where laughter really hurts" (1994, IX). The author's decision to portray AIDS in a way that was at the same time brutally honest and undeniably funny is what embodies the subversive spirit of his works. While portraying his own experience, Feinberg manages to overturn the stereotypes around PWAs. Sick people are portrayed as subversive, not victims who passively accept their condition: "my doctor assured me that there was a minimal likelihood of any side effects with the initial low dosage given my relatively robust blood counts (robust compared to what? anorectic ballerinas? autopsy reports?)" (Feinberg 1991, 133).

The use of humor represents the most notable creative addition on Feinberg's part to his painfully frank style, as it turns a tragic situation into something else. In his writing, Feinberg shapes a different role for sick gay people, rejecting the stereotype of the passive medical patient that was prevalent until the 1980s. This is a shift that is observable in AIDS writing in general: PWAs are usually portrayed as proactive and angry, repudiating the stigma and shame that the mainstream discourse had attempted to attach to AIDS. Feinberg, however, brings this subversion to a different level, both when it comes to brutal honesty and when it comes to black humor. Novelist Sarah Schulman remarks on the coexistence of these two features in Feinberg's writing, noting how "he wasn't sentimental at the time, when a lot of people were, and that made the books seem acerbic and special. . . . David was so angry it was funny, until it became just pure pain"

(Schulman 2012, chap. 3). His creative contribution to AIDS literature has not so much to do with the events he portrays as it has with the point of view from which he describes them. His use of humor is what changes the experience of the epidemic.

Feinberg's use of black humor is part of a long tradition of laughing at something that is deemed inappropriate. As a gay Jewish New Yorker, Feinberg combines different influences in his style, from camp to Jewish black humor. Some AIDS jokes incorporate references to Jewish humor, like when B.J. wonders whether safe sex imposes any rules resembling kosher dietary laws (Feinberg 1991, 140). Apart from that, Feinberg's identity as a Jewish man represents a crucial component of his style in a broader sense. His use of dark humor echoes the lesson of Jewish black humor on the Holocaust, which started to develop in the Sixties in order to deal with another unspeakable tragedy. In both cases, humor becomes a tool to reframe the experience of a minority group, subverting their role from victims of oppression to storytellers in charge of their narrative: "it sounds too Jewish for my taste. I *am* trying to *forget* my *past*" (3). Drawing from Freud's lesson on humor as an assertion of invulnerability (Freud 1927, 162–63), Feinberg employs black humor as a form of "spiritual resistance among the oppressed" (Lipman 1991, 11). While he is not the only author to employ Jewish black humor in this way, again he brings this dynamic to a new, harsher level. The darkness of the humor, however, should not suggest despair: "who says you're going to die? Think positive!" (Feinberg 1991, 190). On the contrary, dark humor implies resistance. While filled with fury, Feinberg's humor is an expression of optimism because, as argued by Patrick O'Neill, "our laughter may contain many degrees of bitterness and hollowness, mirthlessness and parody and pain, but in the end – we do laugh, and while we laugh there's hope" (O'Neill 2010, 100). The fact that Feinberg makes fun of the darkest of circumstances therefore has to be interpreted as an act of resilience. In his article "The People of the Joke," Elliott Oring identifies three fundamental features in Jewish humor: transcendence, defensiveness, and pathology. Humor is transcendent, Oring writes, because it is "transcending the conditions of despair" (1983, 268) which originate it. This dynamic is also at work in Feinberg's writing and represents a direct influence of Jewish black humor, particularly when dealing with the pervasive presence of death: "somebody died. Actually, two people. I might as well buy condolence cards wholesale" (Feinberg 1991, 154). This dark quality of Feinberg's style, as well as his personal involvement with activist groups such as ACT UP,[2] is what allows us to consider him as an

2 The AIDS Coalition To Unleash Power, founded in 1987 by Larry Kramer, is the most notable activist group to come out of the epidemic and made performative demonstrations a cornerstone of AIDS activism.

activist author, employing not merely his work but the totality of his life in order to affect change. Not only is his work protest writing, he himself is a protester (cf. Weir 2010, 33). The closeness of Feinberg to the events he portrays is behind not just his decision to employ black humor, but also what prompts him to adopt an alter ego – another filter he tries to put between himself and the pain brought by AIDS.

Feinberg, B.J. Rosenthal, and the ghost of the epidemic past

In an analysis on autobiographical influences, the relationship between author and character has a central significance. As aforementioned, throughout his fiction production Feinberg recurs to the filter of an alter ego as the main character. B.J.'s presence can be explained as a literary device employed in order to portray the experience of AIDS from a distance, providing at the same time a sense of objectivity about the story and a sense of protection for the author himself while he is offering such a personal and traumatic account. B.J., however, is explicitly Feinberg's alter ego, as they share numerous characteristics both in the material and the intellectual sense. Therefore, the filter is quite thin. Feinberg's fiction accurately tracks the progress of his worries, his diagnosis and illness (cf. Weir 2010, 17): *Eighty-Sixed* was written before him being diagnosed, and *Spontaneous Combustion* deals with the diagnosis. Furthermore, like many in AIDS writing, Feinberg portrays not merely the experience of illness, but also the dynamics of a community where people suddenly had to take care of each other. While caring for Bob, B.J. finds himself constantly thinking about the possibility of having gotten infected himself. The fear of 'having it,' and the fear of people he loves having it, fills his thoughts every day. In the first part of *Spontaneous Combustion*, B.J. obsesses over all his possible symptoms (Feinberg 1991, 18) and ultimately decides to take an HIV test. *Queer and Loathing* also is representative of the time it was written in, as the author lets go of B.J.'s protective mask and directly addresses the reader, knowingly leaving the book as a literary testament. Writing that it was time to come clean (Feinberg 1994, XI), Feinberg still reclaims his right to change a few details: "as I dance at the border of fiction and fact, this is as close to the truth as I can get" (XII). As we will see below, the act of testimony is for him, like for many of his contemporaries, a crucial purpose behind the decision to write. In the introduction to *Queer and Loathing*, as part of his decision to 'come clean,' he explicitly states the intent of his work:

> I've tried to capture what is to me a painfully obvious reality that is rarely written about: what it is like for a gay man to live in the epicenter of the AIDS epidemic; what it is like to be HIV-positive in the nineties; what it is like to outlive one therapist, two dentists, two doctors, and one gastroenterologist. At times I'm shocked when I feel that no one else is documenting this personal history without transmuting it to some metaphoric plane. (XI)

It is important to note that the decision to drop the mask is probably prompted by a renewed sense of urgency due to his terminal condition. The novels, however, still represent a pretty reliable account of the author's life.

The autobiographical element in Feinberg's writing, as it pertains to the narrator's voice as well as to the events depicted, is pervasive. B.J.'s "healthy sense of nihilism" (Bradfield 1991), as Scott Bradfield calls it in his review of *Spontaneous Combustion*, mirrors Feinberg's: the author portrays himself as furious as he was in real life. Furthermore, as we will see, since many of those around him were AIDS authors as well, several events described in Feinberg's books are also referenced to in other texts. It is therefore interesting to observe the balance between author and character and its subsequent deconstruction on Feinberg's part. Autofiction may have influenced this dynamic. In her book *The Story of 'Me,'* Marjorie Worthington writes:

> for decades scholars of life writing have struggled to categorize texts that purposely straddle the line between fiction and nonfiction. Sidonie Smith and Julia Watson point out that "many contemporary writers deliberately blur the boundary between life writing and the kinds of stories told in the first-person novel that some call 'faction,' others 'autofiction'" (2010, 10). Such works have also been called 'nonfictional novels,' 'autobiographical novels,' 'factual fictions,' and 'fictional autobiographies' (Adams 1990, 6). I have been using the term 'autofiction' to refer to texts that, through their author-character, are somewhat referential but that also feature decidedly fictional components. Because of the onomastic author-character, autofiction is often similar to, and sometimes initially mistaken for, memoir. (2018, chap. 5)

While *Eighty-Sixed* and *Spontaneous Combustion* could not themselves be assimilated into the genre of autofiction (the author and the character do not share the same name) and would probably fall into a space between autofiction and autobiographical novel, with *Queer and Loathing* the filter between Feinberg and B.J. falls but the fiction is also no longer present. The author himself, though, declares the influence of autofiction on his writing by referencing Hunter S. Thompson's work as an inspiration for *Queer and Loathing* (cf. Feinberg 1994, 3). When it comes to the novels, on the other hand, the only element preventing us from calling his work autofiction lies in the fact that author and character do not share a name, despite the fact that B.J. is clearly a "writer-character." Worthington argues that

> much like the New Journalist, the trauma fiction writer-character, by making manifest his particular position, demonstrates the impossibility of objectivity. This extreme subjectivity provides a clearer and seemingly more truthful depiction of the traumatic event than an objective account ever could. Thus the biographical connection evoked by the eponymous author-character enables trauma autofiction to explore issues of identity and reality, in addition to plumbing the narrative possibilities and challenges inherent in narrativizing trauma and the potential for truth-telling narratives more generally. (2018, chap. 4)

Truth-telling and, of course, trauma are crucial to AIDS writing, testimonial writing, and most life writing in general. Quoting Cathy Caruth, Worthington states that "the person suffering through a trauma does not or cannot understand the event as it happens, but afterward repeatedly returns viscerally and often unwillingly to that event in his/her memory" (2018, chap. 4): we cannot therefore consider Feinberg's work as trauma fiction as such, despite the fact that he is doubtlessly writing about traumatic events. His premature death does not allow him to develop post-traumatic experiences as he is still well into the trauma at the time of both his writing and his demise, which he knows is coming. His work, then, has to be understood as testimony, which is a central theme in AIDS texts in general. The decision to employ this kind of life writing can be interpreted in connection to the activist spirit in Feinberg's work: Graham Matthews writes that "autofictions are able to heighten emotional engagement while signaling the artfulness of their construction" (2018, 141), and emotional engagement is clearly important when the purpose behind writing lies in portraying an experience "that no one else is documenting." Hywel Dix also notes how "many works of autofiction have been written in the aftermath of some kind of traumatic experience" (2018, 4), as it is the case with AIDS writing. Since Feinberg simultaneously employs different literary devices and embraces different genres to bear witness to the AIDS experience, the tension between fiction and non-fiction in his writing is destined to remain unresolved.

An important part of the argument relating to the balance between fiction and non-fiction in Feinberg's work, and another common trait between the author and B.J., has to do with the obsession with lists. Feinberg compulsively compiled lists about anything and everything, a fixation he admits to in *Queer and Loathing* (cf. Feinberg 1994, 61). This habit echoes Gunnthorunn Gudmundsdottir's argument that life writing represents "fertile ground for experimental writing" (Gudmundsdottir 2003, 1) since lists are prominently featured in Feinberg's fiction, making it a hybrid genre.[3] In *Eighty-Sixed*, for example, the author alternates chapters with non-fiction sections, which often consist of lists. Throughout the book, it is never clear whether these intermissions are written in Feinberg's

3 About life writing as experimentation, cf. Dix (2018, 3).

or B.J.'s voice, thus feeding into the idea that Feinberg's work lies in a hazy territory between fiction and non-fiction. The lists are featured prominently in the entirety of his work, and carry out different functions. First of all, lists can provide a humorous way to talk about the darkest aspects of the epidemic, like "Notes on Death" or "Ethical Suicide Alternatives."[4] This type of list often combines different kinds of intertextual, pop-culture references. Max Saunders remarks that "life writing is fundamentally intertextual" (Saunders 2010, 5), and Marlene Kadar refers to the coexistence of "both 'high' and 'non-high' forms" (Kadar 1992, 11) in life writing. Feinberg's humorous lists offer a great opportunity to observe this combination in his work. Secondly, Feinberg uses lists to detail aspects of the physical, medical reality of having AIDS, from listing symptoms to giving safe sex tips,[5] to a list on "How to Visit Someone in the Hospital with a Terminal Disease."[6] Finally, Feinberg also employs lists to recount some of the most personal, human aspects of being terminally ill, as for example in the list "Regrets."[7] No matter the nuance, however, lists have a specific function in his work – that is, to once again regain control over the virus, as "the meaning of lists is inextricably linked with questions of order" (von Contzen 2021, 40), and humorously rant about it. The elements included in Feinberg's lists are often universal components of any PWA's life: Lauren Berlant notes how "in the contemporary consumer public . . . all sorts of narratives are read as autobiographies of collective experience. The personal is the general. Publics presume intimacy" (Berlant 2008, VII). We find, in Feinberg's work, a specific collective employment of a profoundly personal experience, a form of what Berlant calls 'intimate publics' which, she argues, can have political implications when employed in a time of crisis (cf. Berlant 2011, 184). While this dynamic is present, to a degree, in AIDS literature in general, Feinberg's raw style represents a particularly interesting incarnation of the political relevance of intimacy in AIDS life writing.

The power of Feinberg's figure disrupting and collapsing any distance between life and fiction managed to live on after his death – not only because *Queer and Loathing* was published posthumously. David Feinberg returns on the page in John Weir's 2006 novel *What I Did Wrong*. Weir belongs to the same generation as Feinberg and was his best friend, and in this book he deals with the trauma of Feinberg's death. The novel echoes Feinberg's style, and portrays Feinberg's last days as Weir takes care of him – which makes the book emblematic when it comes to post-traumatic writing and lack of distance between author and charac-

4 Both lists are included in *Queer and Loathing*.
5 Both lists are included in *Eighty-Sixed*.
6 The list is included in *Queer and Loathing*.
7 The list is included in *Queer and Loathing*.

ter. In Weir's book, like in Feinberg's writing, we have alter egos: Weir's is called Tom and Feinberg's is Zack. Apart from the names, every detail seems to be accurate: Weir himself talked about his novel as "posing as fiction."[8] Moreover, his accounts correspond to Feinberg's or to other authors', such as Sarah Schulman. In Weir's novel, again, Zack behaves like Feinberg, for example by documenting everything while he is terminally ill (cf. Weir 2006, 27). He also shares with him some practical details, such as his address. Specific episodes from Feinberg's life, like a disruptive and hilarious visit at Macy's a few days before his death, are included in Weir's book.[9] Therefore, as Feinberg's life remains subject of narration even after his demise, in a way his life writing does not stop. Weir recounts his funeral, a religious rite at the parents' insistence, which turned out to be as painful and hilarious as the rest of Feinberg's accounts: as the parents were late,

> Zack was naked in a pine box loaded with ice to keep the body 'fresh.' The coffin was sweating as the ice was slowly dissolving. We had a bucket under the corner to catch the drips. I thought of Zack floating inside this cocktail like a lemon wedge. All the mourners were sitting quietly in the pews listening to the *drip, drop, drip* into the bucket. (Weir 2006, 204)

Thanks to Weir's novel we can also identify intertextual details that allow us to put Feinberg's writing into a different context. For example, the author closes *Eighty-Sixed* with a deeply moving passage, a metaphor reflecting on the impact of the epidemic: "it begins as a gentle rain. Just a drop, for each illness, each death. And with each passing day it gets worse. Now a downpour. Now a torrent. And there is no likelihood of its ever ending" (Feinberg 1989, 326). This passage stands out in Feinberg's work, as it expresses a rare moment of somber reflection. In *What I Did Wrong*, however, Weir includes a scene set in a literary workshop for PWAs, where a character named Bobby reads a similar text – with a twist:

> He came to his first meeting with a pillow, arranged it on his chair, sat down, and immediately started reading one of his poems which seems about rainfall. He'd list everything the raindrops touched or passed as they fell – cirrus clouds, custard fog, the blue dome of the Chrysler Building, kids on the Grand Concourse cupping their hands around a joint, the rain-pocked tinfoil surface of the East River, sunk cars, inner tubes, silverfish, silt, and then the core of the earth. And we'd say, "Bobby, that was beautiful, what was it about?" And he'd say, "It's about what it's like to shit with rectal tumors." (Weir 2006, 49)

8 Cf. *Transmissions: The Literature of AIDS*, The New School, 2011.https://www.youtube.com/watch?v=avwJZTUGAvI&t=4499s.

9 Cf. Weir (2006, 179–81). Schulman includes both this and the following episode in *The Gentrification of the Mind.*

We can therefore see how Weir embodies Feinberg's subversive, politically incorrect spirit in the novel, carrying on the portrayal of the darkest and yet hilariously recounted aspects of the epidemic. In this sense, we could argue that *What I Did Wrong* is haunted by Feinberg's presence, as some of the passages echo Feinberg's style more than they do Weir's. What we can infer from this dynamic is that there is not merely a lack of distance between Feinberg and B.J., there is also a lack of distance between Feinberg and Zack. Ultimately, there might be a lack of distance between Feinberg and the written word. Worthington writes that "autofiction will not allow us to forget or ignore the authorial presence inherent in any text; autofictional authors refuse to die" (2018, coda). In this case, Feinberg *literally* refuses to die: his tormented presence inhabits his own character and his own books, but also manages to haunt Weir's novel after his death.

Conclusion

If we consider the totality of Feinberg's work, we can identify a specific purpose behind his writing, and that is to provide a counternarrative to the mainstream portrayal of the gay experience during the AIDS epidemic. His work follows the tradition, outlined by Graham and Lucero, of life writing as a "conscious act of self-invention" (2020, chap. 11) challenging the national narrative. By subverting the trope of passivity of PWAs, and by offering such a personal input to the world he portrays, David B. Feinberg has managed to shape a different space for AIDS and LGBTQ representation, one which transforms the filters – particularly the use of humor – into a magnifying glass. In his work, the humorous and the deeply personal traits do not have the effect of perpetuating his trauma, but rather to exorcize it by reclaiming it, making the PWA experience visible to all who will read it. By portraying his experience in detail, Feinberg forces the reader to carry some of that burden for him. In a gay community which was debating how appropriate the use of humor was in dealing with AIDS,[10] Feinberg reclaims the principle that humor can be brutal and, therefore, more effective than somberness.

10 In his essay *Esthetics and Loss*, novelist Edmund White makes an argument to this effect and advises to "avoid humor, because humor seems grotesquely inappropriate to the occasion. A sniggering or wise cracking humor puts the public (indifferent when not uneasy) on cozy terms with what is an unspeakable scandal: death. Humor domesticates terror, lays to rest misgivings that should be intensified. Humor suggests that AIDS is just another calamity to befall mother camp, whereas in truth AIDS is not one more item in a sequence, but a rupture in meaning itself. Humor, like melodrama, is an assertion of bourgeois values; it falsely suggests that AIDS is all in the family. . . . Only a dire gallows humor is acceptable" (White 1994, 216).

Through the brave display of his own life on the page, albeit with a couple of filters but absolutely no frills, Feinberg's counternarrative is one of resilience and subversion. It is a narrative that exists at the core of gay culture, it expresses a spirit akin to what animates Pride every June, opting for a sense of disruptive celebration in spite of the darkness. Feinberg's humorous display of his experience as a PWA manages to be as unruly and effective as that first brick thrown in the early morning of June 28th, 1969 in Greenwich Village. Like a brick, his furious prose mercilessly hits the reader and, through the implication that gay people would have deserved better, David B. Feinberg manages to turn that brick into a symbol of hope.

Works cited

Berlant, Lauren. 2008. *The Female Complaint. The Unfinished Business of Sentimentality in American Culture*. Durham: Duke University Press.

Berlant, Lauren. 2011. Interviewed by Jay Prosser. *Biography* 34 (Winter): 180–87.

Bradfield, Scott. 1991. "Young, Single, and HIV-Positive." *The New York Times* online edition, November 17.

Couser, G. Thomas. 2021. *The Work of Life Writing. Essays and Lectures*. New York: Routledge.

Dix, Hywel. 2018. *Autofiction in English*. Cham: Palgrave Macmillan.

Feinberg, David. 1989. *Eighty-Sixed*. New York: Penguin.

Feinberg, David. 1991. *Spontaneous Combustion*. New York: Penguin.

Feinberg, David. 1994. *Queer and Loathing. Rants and Raves of a Raging AIDS Clone*. New York: Penguin.

Freud, Sigmund. 1927. "Humor." *Standard Edition* 21: 161–66.

Graham, Maryemma, and Lucero, Mercedes. 2020. "Life Writing." In *The Routledge Handbook to the Culture and Media of the Americas*, ed. Wilfried Raussert et al. New York: Routledge, EPUB book.

Gudmundsdottir, Gunnthorunn. 2003. *Borderlines. Autobiography and Fiction in Postmodern Life Writing*. Amsterdam: Rodopi.

Kadar, Marlene. 1992. *Essays on Life Writing. From Genre to Political Practice*. Toronto: University of Toronto Press.

Kushner, Tony. 1994. *Introduction to Queer and Loathing. Rants and Raves of a Raging AIDS Clone*. New York: Penguin.

Lee, Hermione. 2005. *Body Parts*. New York: Random House.

Lipman, Steve. 1991. *Laughter in Hell: The Use of Humor during the Holocaust*. Northvale: Jason Aronson.

Matthews, Graham J. 2018. "Illness Narratives and the Consolations of Autofiction." In *Autofiction in English*, ed. Hywel Dix, 125–43. Cham: Palgrave Macmillan.

Nelson, Emmanuel. 1992. *Introduction to AIDS: The Literary Response*. New York: Twayne Publishers.

O'Neill, Patrick. 2010. "The Comedy of Entropy: The Contexts of Black Humor." In *Dark Humor*, ed. Harold Bloom, 79–103. New York: Infobase.

Oring, Elliott. 1983. "The People of the Joke: On the Conceptualization of a Jewish Humor." *Western Folklore* 42, no. 4: 261–71.

Saunders, Max. 2010. *Self Impression. Life Writing, Autobiografiction, & the Forms of Modern Literature*. Oxford: Oxford Univercity Press.

Schulman, Sarah. 2012. *The Gentrification of the Mind. Witness to a Lost Imagination*. Berkeley: University of California Press, EPUB book.

Texier, Catherine. 1989. "When Sex Was All That Mattered." *The New York Times* online edition, February 26.

von Contzen, Eva. 2021. *Introduction to Lists and Catalogues in Ancient Literature and Beyond*, ed. by Rebecca Laemmle, Cedric Laemmle, and Katharina Wesselmann. Berlin: De Gruyter.

Weir, John. 2006. *What I Did Wrong*. New York: Penguin.

Weir, John. 2010. Interviewed by Sarah Schulman. ACT UP Oral History Project, June 24.

Weir, John. 2011. *Transmissions: The Literature of AIDS*. YouTube, uploaded by The New School, March 12.

White, Edmund. 1994. "Esthetics and Loss." In *The Burning Library: Essays*, ed. by David Bergman, 211–17. New York: A.A. Knopf.

Worthington, Marjorie. 2018. *The Story of 'Me.' Contemporary American Autofiction*. Lincoln: University of Nebraska Press, EPUB book.

Notes on Contributors

Mita Banerjee is professor and chair of American Studies at the Obama Institute for Transnational American Studies at the University of Mainz, Germany. Her research interests include life writing research, whiteness studies, indigenous studies, aging studies, and medical humanities. At the University of Mainz, she currently serves as the director of the Gutenberg Research College (GFK), and she is a member of the advisory board of the German Association for American Studies. From 2014–2023, she was co-speaker of the DFG Research Training Group "Life Sciences, Life Writing: Boundary Experiences of Human Life between Biomedical Explanation and Lived Experience." She is a principal investigator in the Collaborative Research Unit "Humandifferenzierung" (SFB 1482) as well as the Research Group "CORE: Critical Online Reasoning in Higher Education."

Sofie Behluli works as a postdoctoral researcher at the University of Bern, where she researches and teaches for the American section of the English department (chair: Prof. Dr. Gabriele Rippl). She completed her doctorate at the University of Oxford with a thesis on "Women Writing Novels on Art" (2021; supervised by Prof. Dr. Laura Marcus). Her dissertation was fully funded by the Berrow Foundation Scholarship and explored ekphrasis in contemporary women's fictions. Sofie's research interests cover intermediality, life writing, affect studies, conceptions of the 'contemporary' and women's writing. Sofie has published articles in *Women*: A *Cultural Review* (2021) and *Anglia*: *A Journal of English Philology* (2022), and her first monograph is under contract with Oxford University Press.

Anna Ferrari holds a PhD in American Literature from Sapienza, University of Rome and teaches American Literature and English at Sapienza, University of Rome and at the University of Florence as an adjunct professor. Among her research interests are AIDS literature, LGBTQ culture, camp, Jewish humor, pop culture, and representation of the urban space.

Atalie Gerhard (she/her) has lectured at Paderborn University, the University of Tübingen, Potsdam University, and the University of Texas at El Paso and is a doctoral candidate at Saarland University. There, she was a doctoral researcher in the International Research Training Group "Diversity: Mediating Difference in Transcultural Spaces." She was a member of the Emerging Scholars' Forum of the Association for Canadian Studies in German-Speaking Countries. Her research interests include Black and Indigenous resistance and self-representations of women in North American art and literature. She holds a Master of Arts degree in North American Studies as well as a Bachelor of Arts degree in English and American Studies and French Studies from the Friedrich-Alexander-University of Erlangen-Nuremberg.

Astrid Haas is an associate professor in American Literature and Culture at the University of Bergen, Norway. She received her doctorate in English and American Studies from the University of Münster and her postdoctoral degree in North American Literary and Cultural Studies from Bielefeld University (both Germany). Her research interests include transnational North American, Inter-American, and Atlantic studies from the eighteenth to the twenty-first century, the genres of travel writing, drama, and autobiography, the Black and Latinx diasporas, as well as gender and sexuality studies, science studies, and popular culture. Her current major research project is titled "Black Inter-American Mobilities and Autobiography in the Age of Revolutions (1760–1860)."

https://doi.org/10.1515/9783111551968-013

Alfred Hornung is research professor and director of the Obama Institute for Transnational American Studies at the Johannes Gutenberg University of Mainz and an honorary chair professor of Shandong University. His publications are in the field of modernism, postmodernism, ecology, and life writing. He is editor-in-chief of the *Journal of Transnational American Studies* and on the editorial board of several journals, including *Atlantic Studies*, *Contemporary Foreign Literature* (Nanjing), *Life Writing in Europe*, *a/b: Auto/Biography Studies*, and *Life Writing*. He is a founding member of IABA and of IABA-Europe, the recipient of the Bode-Pearson Prize of the American Studies Association and an elected member of Academia Europaea. Among his publications related to life writing are the Chinese translation of *Ecology and Life Writing* (2016), *The Routledge Companion to Transnational American Studies* (2019), and a biography of Al Capone (2021).

Frank Mehring is a professor of American Studies at Radboud University, Nijmegen, Netherlands. His research explores cultural transfer, migration, intermediality, and the role of American music in international contexts. He is a founding member of the EAAS Digital Studies Network and the journal *AmLit-American Literatures*. Frank is the recipient of the EAAS Rob Kroes Award. He serves on the boards of the Netherlands American Studies Association and the European Association for American Studies. Additionally, he is a board member of the EAAS Monograph Series and an Advisory Board member for the *Journal for Polish American Studies*. Frank is deeply committed to public engagement. He has curated concerts featuring Marshall Plan and liberation songs, organized traveling exhibitions, and presented his work on national radio and television.

Wilfried Raussert is professor and chair of North American Studies at Bielefeld University. He is currently chair of the Department of Cultural Studies. From 2009 until 2023 he was director of the International Association of Inter-American Studies. From 2015 until 2018 he was international representative of the American Studies Association (USA). He chaired the BMBF project The Americas as Space of Entanglement (2013–2020) financed by the Federal Ministry of Education and Research in Germany. Currently he is Director of Research at the CALAS Maria Sybilla Merian Center for Advanced Latin American Studies (Guadalajara-Bielefeld). He is the general editor of *fiar forum for inter-american research* journal and of several book series dedicated to InterAmerican Studies (Routledge, WVT/UNO, KIPU). He is on the editorial board of American Studies International. He is also an award-winning photographer with several international photo exhibits, documentary filmmaker with film presentations at global film festivals, and music composer with three music albums to his credit.

Susana Rocha Teixeira is a postdoctoral research associate in the DFG Research Training Group "Authority and Trust" at the Heidelberg Center for American Studies, where she is completing her second book "Re-Imagining the New Negro Renaissance: The Black Renaissance in Baltimore and Beyond." She completed her PhD in American Literature and Culture at Heidelberg University, where she explored the American makeover culture and masculinities. From 2017–2021 she worked as a research associate for the DFG-funded collaborative research project "Practices of Comparing" 1288 at Bielefeld University. She was also a visiting scholar in residence at the History Department at the American University in Washington DC, and at the Krieger School of Arts and Sciences/Center for Advanced Media Studies at Johns Hopkins University. She is a co-editor of *fiar* (the forum for inter-american research) and associated member at Bielefeld University at the SFB 1288. Her research interests as Americanist include makeover culture, masculinities, and American literature and culture from the eighteenth century to the twenty-first century.

Ines E. Veauthier is a university lecturer at Johannes Gutenberg University of Mainz, Germany, School of Translation, Linguistics, and Cultural Studies at the Germersheim campus. The focus of her teaching and research is on literature and culture in the Americas, media studies, and intercultural communication. She is currently involved in a project on gender and translation.

Sabrina Völz is a post-doctoral researcher and senior lecturer in English and American Studies at Leuphana University Lüneburg. Her teaching and research interests focus on ethnic and religious groups in American literature and film. Apart from the article in this volume, Sabrina Völz has published several research essays on ex-Amish memoir. She also teaches life writing to undergraduates and wrote for and co-edited the *American Studies Blog* (http://blog.asjournal.org/) from 2014 to 2023.

Index

https://doi.org/10.1515/9783111551968-014